22.

Animal cell culture

a practical approach

TITLES PUBLISHED IN
THE
PRACTICAL APPROACH
SERIES

Animal cell culture

a practical approach

Edited by

R I Freshney

Cancer Research Campaign, Department of Oncology,
University of Glasgow, 1 Horselethill Road,
Glasgow, G12 9LX, UK

 IRL PRESS
Oxford · Washington DC

IRL Press Limited,
P.O. Box 1,
Eynsham,
Oxford OX8 1JJ,
England

©1986 IRL Press Limited

British Library Cataloguing in Publication Data

Animal cell culture: a practical approach.
—(Practical approach series)
 1. Cell culture
 I. Title II. Freshney,R.I. III. Series
 591'.0724   QH585

ISBN 0-947946-33-0 (softbound)

ISBN 0-947946-62-4 (hardbound)

Cover photograph. Primary culture of cells from whole 12 day mouse embryo after disaggregation in trypsin.

Printed in England by Information Printing, Oxford.

Preface

In recent years the use of animal cell culture has undergone a major expansion from being a purely experimental procedure to become an accepted technological component of many aspects of biological research and commercial exploitation. New developments have arisen to improve both versatility and standardisation and it is the aim of this book to make these developments readily accessible to scientists and technologists in the field. These developments include major advances that have been made in serum-free culture and in the scaling up of animal cell culture to industrial and semi-industrial levels. The importance of cell line characterisation, standardisation and banking are now widely recognised and information is presented to allow the reader both to perform their own banking and to take advantage of central repositories and data banks.

Analytical and preparative fractionation of cell populations is dealt with in two chapters on Flow Cytometry and Centrifugal Elutriation, and analysis at the cellular and sub-cellullar level is described in a chapter on *in situ* hybridisation. The importance of cell interaction and retention of histological structure is stressed in a chapter on organ culture, where most of the standard techniques are described. Finally, a chapter on viability and cytotoxicity measurements has been included to cover the requirements of *in vitro* toxicology and anti-neoplastic drug screening, as well as routine monitoring of culture viability.

Ian Freshney

Contributors

D.Conkie
Beatson Institute for Cancer Research, Wolfson Laboratory for Molecular Pathology, Garscube Estate, Bearsden Road, Glasgow G61 1BD, UK

E.Erba
Istituto di Ricerche Farmacologiche 'Mario Negri', Via Eritrea 62, I-20157 Milano, Italy

R.I.Freshney
Cancer Research Campaign, Department of Oncology, University of Glasgow, 1 Horselethill Road, Glasgow G12 9LX, UK

B.Griffiths
PHLS CAMR, Porton, Salisbury, Wiltshire SP4 0JG, UK

R.J.Hay
American Type Culture Collection, 12301 Parklawn Drive, Rockville, MD 20852, USA

I.Lasnitzki
Strangeways Laboratory, Wort's Causeway, Cambridge CB1 4RN, UK

H.R.Maurer
Freie Universität Berlin, Fachbereich Pharmazie, Institut für Pharmazie (WE1), Königin-Luise-Strasse 2+4, 1000 Berlin 33, FRG

L.Morasca
Istituto di Ricerche Farmacologiche 'Mario Negri', Via Eritrea 62, I-20157 Milano, Italy

A.P.Wilson
Department of Reproductive Pathology, St Mary's Hospital, Hathersage Road, Manchester M13 0JH, UK

Contents

Abbreviations

ATCC	American Type Culture Collection
BSS	balanced salt solution
CAM	chorioallantoic membrane
CSIB	cell source information banks
dCTP	deoxycytidine triphosphate
DGDW	double glass distilled water
dUTP	deoxyuridine triphosphate
EDTA	ethylenediamine tetraacetic acid
HBSS	Hanks' balanced salt solution
HEK cells	human embryonic kidney cells
HLA	human leukocyte antigens
IEF	isoelectric focusing
INT	2-(p-iodophenyl)-3-(p-nitrophenyl)-5 phenyl tetrazolium chloride
LED	light emitting diode
MEM	minimal essential medium
ODR	oxidation-reduction (redox) potential
OTR	oxygen transfer rate
PA	prostate antigen
PBMEM	phosphate-buffered Eagle's MEM
PBS	phosphate-buffered saline
PMT	photomultiplier tubes
RBCs	red blood cells
SSC	standard saline citrate

CHAPTER 1

Introduction: Principles of Sterile Technique and Cell Propagation

R.I.FRESHNEY

1. INTRODUCTION

The culture of animal cells and tissues is now a widely used technique in many different disciplines from the basic sciences of cell and molecular biology to the rapidly evolving applied field of biotechnology. An introduction to the basic procedures is available in many laboratories and frequently features as an integral part of undergraduate study in the biological sciences. Several text books are already available $(1-3)$ to assist the complete novice during his or her introduction to the basic principles of preparation, sterilisation, and cell propagation, so this book will concentrate on certain specialised aspects, many of which are essential for complete understanding and correct utilisation of the technique.

This chapter will review some of the general aspects of cultured cells, their biology, derivation and characterisation, and set out some of the basic assumptions and definitions.

2. BIOLOGY OF CELLS IN CULTURE

2.1 Origin and Characterisation

The list of different cell types which can now be grown in culture is quite extensive, and includes connective tissue elements such as fibroblasts, skeletal tissue (bone and cartilage), skeletal, cardiac and smooth muscle, epithelial tissues (e.g., liver, lung, breast, skin, bladder and kidney), neural cells (glia and neurones, though neurones do not proliferate), endocrine cells (adrenal, pituitary, pancreatic islet cells), melanocytes and many different types of tumour [for further information see (4)].

The use of cell type specific markers (see Chapter 4) has made it possible to determine the lineage from which many of these cultures were derived, but what is not entirely clear, in many cases, is the position of the cells within the lineage. For the cells to proliferate, it is likely that they represent a precursor cell type rather than a fully differentiated cell which would not normally proliferate. However, the population need not be uniform or of a fixed phenotype. Some cultures, e.g., epidermal keratinocytes, contain stem cells, precursor cells and keratinised squames. There is constant renewal from the stem cells, proliferation and maturation in the precursor compartment, and terminal and irreversible differentiation releasing squames into the culture medium. Other cultures, such as fibroblasts, contain a fairly uniform population of proliferating cells at low cell densities ($\sim 10^4$ cells/cm^2) and an equally uniform more differentiated non-proliferating population at high cell densities (10^5 cells/cm^2). This high density

1

population of fibrocyte-like cells can re-enter the cell cycle if the cells are trypsinised or scraped to reduce the cell density or create a free edge.

Nutritional factors like serum or Ca^{2+} ions (5), hormones, cell (6) and matrix interactions (7), in addition to the density of the culture (8), can all affect differentiation and cell proliferation, often inversely. Hence, it is not only essential to define the lineage of cells being used, but also characterise and stabilise the stage of differentiation, by controlling cell density and the nutritional and hormonal environment, to obtain a uniform population of cells.

Because the dynamic properties of cell culture are sometimes difficult to control, and the appropriate cell interactions found *in vivo* difficult to recreate *in vitro*, many people have forsaken the idea of serial propagation in favour of retaining the structural integrity of the original tissue. Such a system is called histotypic or organ culture and is dealt with in Chapter 7. Attempts have also been made to recreate tissue like structures *in vitro* by reaggregating different cell types and culturing at high density as multicellular spheroids (9), perfused multilayers on glass or plastic substrates (10) or floating cultures on collagen (11) or synthetic microporous filters (12).

2.2 Differentiation

As propagation of cell lines requires that the cell number increases continually, culture conditions which have evolved over the years have been selected to favour maximal cell proliferation. It is not surprising that these conditions are not often conducive to cell differentiation where cell growth is severely limited or completely abolished. Those conditions which favour cell proliferation are low cell density, low Ca^{2+} concentration (13) $(100-600~\mu M)$ and the presence of growth factor such as epidermal growth factor (EGF), fibroblast growth factor (FGF), and platelet derived growth factor (PDGF). High cell density ($> 10^5$ cells/cm²), high Ca^{2+} concentration $(300-1500~\mu M)$ and the presence of differentiation inducers [hormones such as hydrocortisone (14), glia maturation factor (15), nerve growth factor (16), retinoids (17) and polar solvents, such as dimethyl sulphoxide (18)], will favour cytostasis and differentiation.

The role of serum in differentiation is not entirely clear and depends on the cell type and medium used. While a low serum concentration promotes differentiation in oligo-dendrocytes (19), a high serum concentration causes squamous differentiation in bronchial epithelium (20). In the latter case, the active principle is a molecule closely resembling or identical to tumour derived growth factor β isolated from platelets. The use of defined media will hopefully help to resolve this question.

The establishment of the correct polarity and cell shape may also be important, particularly in epithelium. Many workers have shown that growing cells to high density on a floating collagen gel allows matrix interaction, access to medium on both sides, the possibility of establishing correct polarity with respect to the basement membrane and the adoption of the correct cell shape due to the plasticity of the substrate (21).

Different conditions are required, therefore, for propagation and differentiation and hence an experimental protocol may require a growth phase to increase cell number and allow for replicate samples, followed by a non-growth maturation phase to allow for increased expression of differentiated functions.

Table 1. Respective Advantages of Cell and Organ Culture.

Organ culture	Cell culture
Histology	Propagation and expansion
Differentiation	Cloning, selection and purification
Cell interaction, homotypic and	
heterotypic	Characterisation and preservation
Matrix interaction	Replicate sampling and quantitation

3. CHOICE OF MATERIALS

The primary determinant in selecting a tissue or cell line for further study is the nature of the observations that have to be carried out. General cellular processes such as DNA synthesis, membrane permeability, or determination of cytotoxicity may be feasible with any cell type, while the study of specialised properties such as myotube fusion, antibody production or regulation of urea cycle enzymes will require cell types expressing these specific functions.

3.1 Organ Culture or Cell Culture?

Originally tissue culture was regarded as the culture of whole fragments of explanted tissue with the assumption that histological integrity was maintained, at least in part. Now 'tissue culture' has become a generic term and encompasses organ culture, where a small fragment of tissue or whole embryonic organ is explanted to retain tissue architecture, and cell culture where the tissue is dispersed mechanically or enzymatically or by spontaneous migration from an explant and may be propagated as a cell suspension or attached monolayer.

In adopting a particular type of culture the following points should be taken into account. Organ culture (see Chapter 7) will preserve cell interaction, retain histological and biochemical differentiation for longer, and, after the initial trauma of explantation and some central necrosis will generally remain in a non-growing steady state for a period of several days and even weeks. They cannot be propagated, generally incur greater experimental variation between replicates, and tend to be more difficult to use for quantitative determinations due to minor variations in geometry and constitution.

Cell cultures on the other hand, are usually devoid of structural organisation, have lost their histotypic architecture and often the biochemical properties associated with it, and generally do not achieve a steady state unless special conditions are employed. They can, however, be propagated and hence expanded and divided into identical replicates, they can be characterised and a defined cell population preserved by freezing, and they can be purified phenotypically by growth in selective media (Chapter 2), physical cell separation (Chapters 5,6) or cloning (Chapters 4,8) and genotypically to give a characterised cell strain with considerable uniformity.

These properties are summarised in *Table 1*.

3.2 Source of Tissue

3.2.1 *Embryo or Adult?*

In general, cultures derived from embryonic tissues will survive and grow better than those from the adult. This presumably reflects the lower level of specialisation and presence of replicating precursor or stem cells in the embryo. Adult tissues will usually have a lower growth fraction and a high proportion of non-replicating specialised cells, often within a more structured, and less readily disaggregated, extracellular matrix. Initiation and propagation are more difficult, and the lifespan of the culture often shorter.

Embryonic or foetal tissue has many practical advantages, but it must always be remembered that in some instances the cells will be different from adult cells and it cannot be assumed that they will mature into adult-type cells unless this can be confirmed by appropriate characterisation.

Examples of widely used embryonic cell lines are the various 3T3 lines (mouse embryo fibroblasts) and MRC-5 and other human foetal lung fibroblasts. Mesodermally derived cells (fibroblasts, endothelium, myoblasts) are also easier to culture than epithelium, neurones or endocrine tissue but this may reflect the extensive use of fibroblast cultures during the early years of the development of culture media together with the response of mesodermally-derived cells to mitogenic factors present in serum. A number of new selective media have now been designed for epithelial and other cell types (see Chapter 2) and with some of these it has been shown that serum is inhibitory to growth and may promote differentiation (20).

3.2.2 *Normal or Neoplastic?*

Normal tissue usually gives rise to cultures with a finite lifespan while cultures from tumours can give continuous cell lines (see below), although there are several examples of continuous cell lines (MDCK dog kidney, 3T3 fibroblasts) which are non-tumorigenic.

Normal cells will generally grow as an undifferentiated stem cell or precursor cell and the onset of differentiation is accompanied by a cessation in cell proliferation which may be permanent. Some normal cells, e.g., fibrocytes or endothelium, are able to differentiate and still dedifferentiate and resume proliferation and in turn redifferentiate, while others, e.g., squamous epithelium and many haemopoietic cells, once initiated into differentiation are incapable of resuming proliferation.

Cells cultured from neoplasms, however, can express at least partial differentiation, e.g., B16 mouse melanoma, while retaining the capacity to divide. Many studies of differentiation have taken advantage of this fact and used differentiated tumours such as the minimal deviation hepatomas of the rat (22) and human and rodent neuroblastomas (23), although whether this can be taken as normal differentiation is always in doubt.

Tumour tissue can often be passaged in the syngeneic host, providing a cheap and simple method of producing large numbers of cells, albeit with lower purity. Where the natural host is not available, tumours can also be propagated in athymic mice with greater difficulty but similar advantages.

Many other differences between normal and neoplastic cells are similar to those between finite and continuous cell lines (see below) and indeed the importance of immortalisation in neoplastic transformation has been recognised (24).

3.3 **Subculture**

Freshly isolated cultures are known as *primary cultures* until they are passaged or subcultured. They are usually heterogeneous, and have a low growth fraction, but are more representative of the cell types in the tissue from which they were derived and in the expression of tissue specific properties. Subculture allows the expansion of the culture (it is now known as a *cell line*), the possibility of cloning (see Chapters 4 and 8), characterisation and preservation (Chapter 4) and greater uniformity but may cause a loss of specialised cells and differentiated properties unless care is taken to select out the correct lineage and preserve or reinduce differentiated properties (see below).

The greatest advantage of subculturing a primary culture into a cell line is the provision of large amounts of consistent material suitable for prolonged use.

3.3.1 *Finite or Continuous Cell Lines?*

After several subcultures a cell line will either die out (*finite cell line*) or 'transform' to become a *continuous cell line*. It is not clear in all cases whether the stem line of a continuous culture pre-exists masked by the finite population or arises during serial propagation. Because of the time taken for such cell lines to appear (often several months) and the differences in their properties it has been assumed that a mutational event (chromosomal rearrangement, translocation, partial or total non-disjunction, or point mutation) occurs, but the pre-existence of immortalised cells, particularly in cultures from neoplasms, cannot be excluded.

The appearance of a continuous cell line is usually marked by an alteration in cytomorphology (smaller cell size, less adherent, more rounded, higher nucleo:cytoplasmic ratio), an increase in growth rate (population doubling time decreases from 36−48 h to 12−36 h), a reduction in serum dependence, an increase in cloning efficiency, a reduction in anchorage dependence (i.e., an increased ability to proliferate in suspension as a liquid culture or cloned in agar), an increase in heteroploidy (chromosomal variation between cells) and aneuploidy (divergence from the donor, euploid, karyotype) and an increase in tumorigenicity. The resemblance between spontaneous *in vitro* transformation and malignant transformation is obvious but nevertheless the two are not necessarily identical although they have much in common. Normal cells can 'transform' to become continuous cell lines without becoming malignant and malignant tumours can give rise to cultures which 'transform' and become more (or even less) tumorigenic but acquire the other properties listed above.

The advantages of continuous cell lines are their greater growth rates to higher cell densities and resultant higher yield, their lower serum requirement and general ease of maintenance in simple media, and their ability to grow in suspension. Their disadvantages include greater chromosomal instability, divergence from the donor phenotype, and loss of tissue specific markers.

3.3.2 *Propagation in Suspension*

Most cultures, including primaries, are propagated as a monolayer, anchored to a glass or plastic substrate. Some cultures, principally transformed cells, haemopoietic cells and ascites tumours can be propagated in suspension. This has the advantage of simpler propagation (subculture only requires dilution, no trypsinisation), no requirement for increasing surface area with increasing bulk, ease of harvesting and the possibility of achieving a 'steady-state' culture if required (see Chapter 3).

3.4 **Selection of Medium**

Regrettably the choice of medium is still often quite empirical. What was used previously by others for the same cells, or what is currently being used in your own laboratory for different cells often dictates the choice of medium and serum. For continuous cell lines it may not matter a great deal as long as the conditions are consistent but for specialised cell types, primary cultures, and growth in the absence of serum, the choice is more critical. This subject is dealt with in more detail in Chapter 2.

There are two major advantages of using more sophisticated media in the absence of serum: they may be selective for particular types of cell, and the isolation of purified products is easier in the absence of serum.

Nevertheless, culture in the presence of serum is still easier and often, surprisingly, no more expensive, though undoubtedly less controlled. Two major determinants still regulate the use of serum-free media (a) they are only gradually becoming available commercially (many people do not have the time, facilities or inclination to make up their own), (b) requirements for serum-free media are more cell-type specific. Serum will cover many inadequacies revealed in its absence. This problem may be particularly acute when culturing tumour cells where cell line variability may require modifications for cell lines from individual tumours.

In the final analysis the choice is often still empirical. Read the literature and determine which medium has been used previously. If several different media have been used (and this is often the case) test them all for yourself, with a few others added to the series if you wish. Test growth (doubling time and saturation density) (see below), cloning efficiency (Chapter 8), and expression of specific properties (differentiation, viral propagation, cell products, etc.). The choice of medium may not be the same in each case, e.g., differentiation of lung epithelium will proceed in serum, propagation is better without. If possible include one or more serum-free media in your panel supplemented with growth factors, hormones, and trace elements as required (see Chapter 2).

Once a type of medium has been selected try to keep this constant for as long as possible. Similarly if serum has to be used select a batch by testing samples from commercial suppliers and reserve enough to last 6 months to one year, before replacing it with another pretested batch.

3.5 **Gas Phase**

The gas phase is determined (a) by the type of medium, (b) by whether the culture vessel is open (Petri dishes, multiwell plates) or sealed (flasks, bottles), and (c) the amount of buffering required. Several variables are at play but one major rule predominates and three basic conditions can be described. The rule is that the bicarbonate concentrate and CO_2 tension must be in equilibrium. The three conditions can be summarised as follows:

	$[HCO_3']$	Gas phase	Hepes
1. Low buffering capacity, sealed flask	4 mM	Air	—
2. Moderate buffering capacity, open or sealed vessel	23 mM	5% CO_2	—
3. High buffering capacity, open or sealed vessel	8 mM	2% CO_2	20 mM

It should be remembered that CO_2/HCO_3' are essential to most cells, so a flask or dish cannot be vented without providing CO_2 in the atmosphere.

Prepare medium to about pH 7.1 or 7.2 at room temperature, incubate with the correct CO_2 tension for at least 0.5 h in a shallow dish and check that the pH stabilises at pH 7.4. Adjust the sterile N HCl or N NaOH if necessary.

Oxygen tension is usually maintained at atmospheric but variations have been described, e.g., elevated for organ culture (see Chapter 7) and reduced for cloning melanoma (see Chapter 8).

3.6 Substrate

The nature of substrate is determined largely by the type of cell and the use to which it will be put. Polystyrene which has been treated to make it wettable and give it a net negative charge is now used almost universally. In special cases (culture of neurones, muscle cells and some epithelial cultures) the plastic is precoated with gelatin, collagen or polylysine to give a net positive charge. Glass may also be used but must be washed carefully with a non-toxic detergent (see below).

Culture vessels vary in size from Teraskia multiwell plates (~ 1 mm² surface area, $5 - 10 \mu l$ medium volume) microtitration plates (~ 30 mm²/$100 - 200 \mu l$) up through a range of dishes and flasks to 180 cm² and roller bottles and multisurface propagators (see Chapter 3) for large scale culture. The major determinants are the number of cells required ($5 \times 10^4 - 10^5$/cm² maximum for most transformed cells), the number of replicates (96 in a microtitration plate) and the times of sampling; a 24 well plate is good for a large number of replicates for simultaneous sampling but individual tubes or bottles are preferable where sampling is carried out at different times. Petri dishes are cheaper than flasks and good for subsequent processing, e.g., staining or extractions. Flasks can be sealed and do not need a humid, CO_2 incubator and give better protection against contamination.

For suspension cultures, volume is the main determinant. Sparging and agitation may be necessary when the volume is higher (see Chapter 3).

4. PREPARATION

4.1 Substrate

Many laboratories now utilise disposable plastics as substrates for tissue culture. They are optically clear, prepared for tissue culture use by modification of the plastic to make it wettable and suitable for cell attachment, and come already sterilised for use. On the whole they are convenient and provide a source of reproducible vessels for both routine and experimental work. However, they are disposable and as such are expensive to use.

4.1.1 *Washing Glassware*

For routine passage of continuous cell lines and many finite cell lines, glass is perfectly adequate and cheaper to use provided you have the glass washing facilities and can control the quality of washing. A non-toxic detergent must be used, the glassware given a thorough soak, preferably overnight and then thoroughly rinsed in tap water followed by deionised or distilled water. It is assumed that all glassware specified in each chapter meets this specification or better (see also Chapter 7).

4.1.2 *Sterilising Glassware*

Although plastics come already sterilised, glassware must be sterilised before use. All the protocols described in subsequent chapters assume that all glassware is sterile. Many laboratories routinely sterilise all glassware, regardless of whether it will be used directly for culture or not, to avoid possible confusion between sterile and non-sterile stocks.

Sterilisation is best done by dry heat (160°C for one hour) in containers or foil covered, with screw caps autoclaved separately.

Periodically it may be necessary to check the quality of your glassware, the wash-up process, or the choice of detergent. This is best done by washing in the usual way, checking the glassware by eye, drying and sterilising the glass and cloning monolayer cells on the glass (see Chapters 4 and 8), preferably in a low serum concentration (2−5%) or serum free (see Chapter 2).

4.2 **Medium**

Most of the commonly used media are available commercially (see Appendix I) but for special formulations (see Chapter 4) or additions it may be necessary to prepare and sterilise your own. In general, stable solutions (water, salts, and media supplements such as tryptose or peptone) may be autoclaved at 121°C (1 atmosphere above ambient), while labile solutions (media, trypsin and serum) must be filtered through a 0.2 μm porosity membrane filter (Millipore, Sartorius, Gelman, Pall).

Where an automatic autoclave is used care must be taken to ensure that the timing of the run is determined by the temperature of the load and not just by drain or chamber temperature or pressure which will rise much faster than the load.

Sterility testing (see Chapter 4) should be carried out on samples of each filtrate.

4.3 **Cells**

4.3.1 *Primary Culture*

When cells are isolated from a tissue, grown *in vitro*, and before subculture, they are regarded as a primary culture. Primary cultures lack many of the cells, present in the original tissue, which were unable to attach and survive *in vitro*. Furthermore, if the culture proliferates, then any non-dividing or slow growing cells will be diluted out. Hence it may be necessary at this stage to select specific cell types by cloning, selective culture (Chapter 2) or physical cell separation (Chapters 5 and 6).

The first step in preparing a primary culture is sterile dissection followed by mechanical or enzymatic disaggregation. The tissue may simply be chopped to around 1 mm³ and the pieces attached to a dish by their own adhesiveness or by scratching the dish, or by using clotted plasma (1). In these cases cells will grow out from the fragment and may be used or subcultured. The fragment of tissue, or explant as it is called, may be transferred to a fresh dish or the outgrowth trypsinised to leave the explant and a new outgrowth generated.

The trypsinised cells from the outgrowth are reseeded in a fresh vessel and become a secondary culture, and the culture is now technically a cell line.

Primary cultures can also be generated by disaggregating tissue in enzymes such as trypsin (0.25% crude or 0.01 − 0.05% pure) or collagenase (200 − 2000 units/ml, crude) and the cell suspension allowed to settle on to, adhere, and spread out on a glass or plastic substrate (1). This type of culture gives a higher yield of cells though it can be more selective as only certain cells will survive dissociation. In practice, many successful primary cultures are generated using enzymes such as collagenase to reduce the tissue, particularly epithelium, to small clusters of cells which are then allowed to attach and grow out (25).

For methods of primary culture see (1).

4.3.2 *Subculture*

A monolayer culture may be transferred to a second vessel and diluted by dissociating the monolayer in trypsin (suspension cultures only need to be diluted). This is best done by rinsing the monolayer with PBS or PBS containing 1 mM EDTA, adding cold trypsin (0.25% crude or 0.01 − 0.05% pure) for 30 sec, removing the trypsin and incubating for approximately 15 min. Cells are then resuspended in medium, counted and reseeded.

4.3.3 *Growth Curve*

When cells are seeded into a flask they enter a lag period of 2 − 24 h, followed by

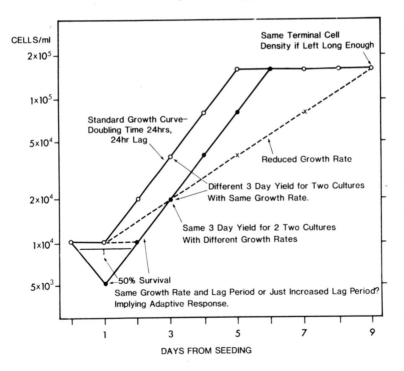

Figure 1. Hypothetical growth curves, calculated from a seeding concentration of 1×10^4 cells/ml, emphasising the need to examine the entire curve and not just count cells at one timepoint.

a period of exponential growth (the 'log phase') and finally enter a period of reduced or no growth after they become confluent ('plateau phase') (*Figure 1*). These phases are characteristic for each cell line and give rise to reproducible measurements — the length of the lag period, the population doubling time in mid-log phase and the saturation density at plateau, given that the environmental conditions are kept constant.

The determination of the growth cycle is important in designing routine subculture and experimental protocols. Cell behaviour and biochemistry changes significantly at each phase and it is therefore essential to control the stage of the growth cycle when drugs or reagents are added and cells harvested. The shape of the growth curve can also give information on the reproductive potential of the culture (*Figure 1*) but it is generally recognised that the analysis of clonal growth is easier (26), and less prone to ambiguity and misinterpretation.

Further discussion of cell growth and viability are presented in Chapter 8.

4.3.4 *Feeding*

Some rapidly growing cultures, e.g., continuous cell lines like HeLa, will require a medium change after 3 – 4 days. This is usually indicated by a fall to below pH 7.0.

4.3.5 *Contamination*

The problem of microbial contamination has been greatly reduced by the use of antibiotics and laminar air flow cabinets. However, cultures should be maintained in the absence of antibiotics whenever possible so that chronic, cryptic contaminations are not harboured.

Check frequently for contamination by looking for a rapid change in pH (usually a fall but some fungi can increase the pH), cloudiness in the medium, extracellular granularity under the microscope, or any unidentified material floating in the medium. If a contamination is detected, discard the flask unopened and autoclave.

If in doubt, remove a sample and examine by phase microscopy, Gram's stain or by standard microbiological techniques (see Chapter 4).

Mycoplasma. Cultures can become infected by mycoplasma from media, sera, trypsin or the operator. They are not visible to the naked eye, and, while they can affect cell growth, their presence is often not obvious. It is important to test for mycoplasma at regular intervals (every 1 – 3 months) as they can seriously affect cellular biochemistry, antigenicity and growth characteristics. Several tests have been proposed but the fluorescent DNA stain technique of Chen is the most widely used (see Chapter 4).

4.3.6 *Cross Contamination*

This problem is dealt with in detail in Chapter 4. Its severity is often underrated and consequently it still occurs with a higher frequency than many people would admit. To avoid cross contamination, do not share bottles of media or reagents between cell lines and do not return a pipette which has been in a flask or bottle with cells back to the medium bottle.

5. REFERENCES

1. Freshney,R.I. (1983) *Culture of Animal Cells: Manual of Basic Technique*, Alan Liss Inc., New York.
2. Paul,J. (1975) *Cell and Tissue Culture*, Vth edition, Livingstone, Edinburgh.
3. Adams,R.L.P. (1980) *Laboratory Techniques in Biochemistry and Molecular Biology — Cell Culture for Biochemists*, Work,T.S. and Burdon,R.H. (eds.), Elsevier, Amsterdam.
4. Harris,C.C., Trump,B.F. and Stoner,G.D. (1980) *Normal Human Tissue and Cell Culture, Methods in Cell Biology*, Vol. **21**, Academic Press, New York.
5. Wakelam,M.J.O. (1985) *Biochem.J.*, **228**, 1.
6. Cunha,G.R. and Chung,L.W.K. (1981) *J. Steroid Biochem.*, **14**, 1317.
7. Kemp,R.B. and Hinchliffe,J.R. (1984) *Matrices and Cell Differentiation, Progress in Clinical and Biological Research*, Vol. **15**, Alan Liss, Inc, New York.
8. Frame,M.C., Freshney,R.I., Vaughan,P.F.T., Graham,D.I. and Shaw,R. (1984) *Br. J. Cancer*, **49**, 269.
9. Freyer,J.P. and Sutherland,R.M. (1980) *Cancer Res.*, **40**, 3956.
10. Kruse,P.F.,Jr. and Miedema,E. (1965) *J. Cell. Biol.*, **27**, 273.
11. Michalopoulos,G. and Pitot,H.C. (1975) *Fed. Proc.* **34**, 826.
12. Chambard,M,. Verrier,B., Gabrion,J. and Mauchamp,J. (1983) *J. Cell. Biol.*, **96**, 1172.
13. Peehl,D.M. and Ham,R.G. (1980) *In Vitro*, **16**, 526.
14. Piddington,R. and Moscona,A.A. (1967) *Biochim. Biophys. Acta*, **141**, 429.
15. Kato,T., Fukui,Y., Turriff,D.E., Nakagawa,S., Lim,R., Arnason,B.G.W. and Tanaka,R. (1981) *Brain Res.*, **212**, 393.
16. Levi-Montalcini,R. (1964) *Science*, **143**, 105.
17. Sporn,M.B. and Roberts,A.B. (1983) *Cancer Res.*, **43**, 3034.
18. Friend,C., Scher,W., Holland,J.G. and Sato,T. (1971) *Proc. Natl. Acad. Sci. USA*, **68**, 378.
19. Raff,M.C,. Miller,R.H. and Noble,M. (1983) *Nature*, **303**, 390.
20. Lechner,J.F., McClendon,I.A., Laveck,M.A., Shamsuddin,A.M. and Harris,C.C. (1983) *Cancer Res.*, **43**, 5915.
21. Sattler,C.A., Michalopoulos,G., Sattler,G.L. and Pitot,H.C. (1978) *Cancer Res*, **38**, 1539.
22. Pitot,H., Peraino,C., Morse,P. and Potter,V.R. (1964) *Natl. Cancer Inst. Monographs*, **13**, 229.
23. Littauer,U.Z., Giovanni,M.Y., Glick,M.C. (1979) *Biochem. Biophys. Res. Commun.*, **88**, 933.
24. Land,H., Parada,L.F. and Weinberg,R.A. (1983) *Nature*, **304**, 596.
25. Freshney,R.I. (1972) *Lancet*, **II**, 488.
26. Ham,R.G. and McKeehan,W.L. (1978) *In Vitro*, **14**, 11.

CHAPTER 2

Towards Chemically-defined, Serum-free Media for Mammalian Cell Culture

H.R.MAURER

1. INTRODUCTION

When a cell is removed from its original tissue or organism and placed in culture, the medium must provide all the environmental conditions that the cell has been exposed to *in vivo*; only then will it be able to survive, to proliferate and to differentiate. The extracellular medium must meet the essential requirements for survival and growth (i.e., must provide nutritional, hormonal and stromal factors). Among the biological fluids that proved successful for culturing cells outside the body, serum gained the most wide-spread significance. Although the requirements which lead to the development of complex, chemically defined media were partially elucidated, 5 – 20% serum was and is still usually needed for optimum cell growth. For a variety of reasons it would be of great advantage to eliminate the indefinable serum constituents, thus creating fully chemically defined media. Some of the various approaches and relevant problems have been described previously (1 – 9).

2. ROLE OF SERUM IN CELL CULTURE

2.1 Role of Serum Components

Serum is an extremely complex mixture of many small and large biomolecules with different, physiologically balanced growth-promoting and growth-inhibiting activities. Some of the serum components that have been found to support survival and growth of many mammalian cells in culture are listed in *Table 1*.

The major functions of serum (5) are to provide:
(i)   hormonal factors stimulating cell growth and functions;
(ii)   attachment and spreading factors (biomatrix);
(iii) transport proteins carrying hormones, minerals, lipids, etc.

2.1.1 Growth Factors

Most growth factors are present in the nano and subnanogram per ml range in serum. Some of them are specific for cells at distinct differentiation stages, like the haematopoietic growth factors (colony stimulating factors), some act synergistically, some are not restricted to one cell type. Epidermal growth factor (EGF), for example, promotes the proliferation of fibroblasts, epidermal and glial cells. Moreover, one

Table 1. Major Serum Components Essential for Cell Survival and Growth *in vitro*.

Component	Main functions
Proteins	
Albumin	Carries lipids, hormones, minerals
	Provides osmotic pressure, buffering capacity
Fibronectin	Promotes cell attachment
α_2-Macroglobulin	Inhibits trypsin
Fetuin	Enhances cell attachment
Transferrin	Binds Fe^{2+}
Polypeptides and growth factors	
Insulin	Promotes uptake of glucose and amino acids by cells
Insulin-like growth factor I and II (IGF)	Mitogens
Somatomedin A and C	Mitogens
Multiplication stimulating activity (MSA)	Mitogen
Platelet-derived growth factor (PDGF)	Mitogen for fibroblasts, smooth muscle cells and others
Epidermal growth factor (EGF)	Mitogen
Fibroblast growth factor (FGF)	Mitogen
Endothelial growth factor (ECGF)	Mitogen
Eye-derived growth factor	Mitogen
Peptides	
Glutathione	Redox reactions
Other hormones	
Cortisol (hydrocortisone)	Promotes cell attachment, induces cell differentiation
Oestrogens, androgens	Mitogen
Thyroid hormones (T_3, T_4)	Oxygen consumption, energy metabolism; promotes growth and differentiation of various cells
Lipids	
Linoleic acid	Membrane biosynthesis
Prostaglandins	
Cholesterol	Membrane biosynthesis
Metabolites	
Amino acids	
α-Keto acids (pyruvate)	
Polyamines	Cell proliferation
Minerals	
Fe^{2+}, Zn^{2+}, Cu^{2+}, Mn^{2+}, SeO_3^{2+}, Co^{2+}, VO_3^{-}, $Mo_7O_{24}^{6-}$	Enzyme activation among others

cell type may be stimulated by different growth factors. Thus fibroblasts respond to fibroblast growth factor (FGF), EGF, platelet derived growth factor (PDGF) and somatomedins. At present the physiological role of these numerous factors is not fully understood; for more on growth factors see reference 10.

2.1.2 *Hormones*

Among the hormones, insulin is essential for the growth of nearly all cells in culture. Due to its short half life and sensitivity to inactivation by cysteine, it is usually added at a relatively high concentration. Glucocorticoids (hydrocortisone, dexamethasone) may stimulate or inhibit cells in culture depending on the cell type and cell density; they may modulate cell proliferation by altering responsiveness to growth factors. Some cell lines require other steroid hormones (e.g., oestradiol, testosterone, progesterone) (see *Table 2*). Thyroid hormones [mainly triiodothyronine (T_3)] seem to support the growth of several cell lines (see *Table 2*). (For references to this and the following subheadings, see *Table 2*.)

2.1.3 *Attachment and Spreading Factors*

A major role of serum is to provide attachment and spreading factors, essential components of the extracellular matrix.

Many mammalian cells must first adhere to an adequate substrate, then spread out before they start proliferating and forming a monolayer. The role of the factors (collagen, fibronectin among others) required for attachment and spreading is discussed below (Section 3.2.3).

2.1.4 *Binding Proteins*

Serum also provides several so called binding proteins the function of which is to carry essential low molecular mass factors. Albumin carries vitamins, lipids (fatty acids, cholesterol) and hormones among others. Iron-saturated transferrin is essential for most cells in culture; indeed they possess a specific transferrin-receptor on the surface.

2.1.5 *Lipids*

Serum is a rich source for the various lipids cultured cells generally need for survival and particularly for growth. Cell lines differ in their requirements for essential fatty acids, phospholipids, lecithin and cholesterol. For example optimum fatty acid/cholesterol/albumin balance varies radically for different haematopoietic lineages and for different maturation stages within a given lineage (11). Prostaglandins E and $F_{2\alpha}$ have been found to be involved in cell proliferation, possibly in a synergistic manner with growth factors like EGF.

2.1.6 *Minerals*

Finally, the role of various inorganic tracer elements (Cu, Zn, Co, Mn, Mo, Va, Se) present in serum has been only partially elucidated but many may act as enzyme co-factors. For example, SeO_3^{2-} is required to activate distinct enzymes essential for metabolic detoxification. Selenium may also serve to inactivate free radicals.

2.2 Potential Advantages of using Low-serum and Serum-free Media

Several benefits result from the use of low-serum or serum-free media:
(i)   improved reproducibility between cultures, avoidance of lot-to-lot variations of sera;

Table 2. Low-protein and Serum-free Media for Various Mammalian Cell Types. Concentrations are Given

Cell type	First author	Basal medium	Insulin µg/ml	Transferrin µg/ml	Selenite E-9	EGF ng/ml	FGF ng/ml	Hydrocortisone E-6
Non-transformed, diploid cells								
1. Fibroblasts, human	Ham (8)	MCDB 110	0.95			30		
2. Fibroblasts, chicken	Ham (8)	MCDB 202	20			20		
3. Fibroblasts, sheep	Ham (8)	MCDB 202	10			–		
4. Fibroblasts, human diploid MRC-5	Clark (20)	DME/199 (1:1)	1	25	(15)	10		
5. Lung fibroblast human (WIT-38)	McKeehan (21)	MCDB 104	5	5		100		
6. Keratinocytes, human	Ham (8)	MCDB 153	5			5		1.4
7. Bronchoepi-thelial cells, human	Ham (8)	MCDB 152	5	10		5		0.5
8. Mammary epi-thelial cells, human	Ham (8)	MCDB 170	5	5		25		1.4
9. Chondrocytes, rabbit	Jennings (22)	MCDB 104	1				100	
10. Chondrocytes, articular, rabbit	Adolphe (23)	F12	5	5	4 ng/ml		100	
11. Prostate cells, human	Chapro-niere (24)	RPMI 1640	12	1		10		
12. Testicular epi-thelial cell line mouse (TM4)	Bottenstein (3)	DME/F12 (1:1)	5	5		3		
13. Neuronal cells	Bunge (25)	DME/F12 (1:1)	5	100	30			10 U/ml
14. Astrocytes mouse	Fischer (26)	BME-Earle	10	100	30	10 nM		
15. Brain cells, fetal, rat	Honegger (27)	DME	5	1				0.02
16. Pituitary cells, rat	Lang (28)	DME/F12	5	5	50	40	10	
17. Hypothalamus cells, rat	Lang (28)	DME/F12	5	100	30			8 ng/ml
18. Granulosa cells, rat	Orly (25)	DME/F12 (1:1)	2	5				40 ng/ml

in Computer Style Exponential Notation: 1.0 E-6 means 1.0 x 10^{-6} M = 1 μM.

Dexamethasone E-6	T_3 E-9	Progesterone E-9	Putrescine E-4	Fibronectin μg/ml	BSA mg/ml	Ethanolamine	Phosphoethanolamine	Additional components and comments
1.4								Prostaglandins: PGE_1 2.5 E-8, PGF 2α 2.0 E-6, dithiothreitol 6.5 E-6, glutathione 6.5 E-7, phosphoenolpyruvate 1.0 E-5, liposome B 10 μl/ml (soy bean lecithin 6 ng/ml, cholesterol 3 μg/ml, sphingomyelin 1 μg/ml, vitamin E 0.06 μg/ml, vitamin E acetate 0.2 μg/ml)
–								Liposome B 10 μl/ml (see above)
1.0								Liposome B 30 μl/ml (see above)
50 ng/ml		1	2	2				Fetuin 1 mg/ml, molybdate 5.0 E-10, Hepes 1.0 E-2
55 ng/ml			5					PDGF 1 μg/ml
						1.0 E-4	1.0 E-4	
						5.0 E-7	5.0 E-7	
						1.0 E-4	1.0 E-4	PGE_1 2.5 E-8, prolactin 5 μg/ml
								Liposome B 5 μg/ml (see No. 1)
20 ng/ml			2	1				Bovine growth factor 10 μg/ml, thrombin 1 μg/ml MSA 0.05 μg
0.01								$ZnCl_2$ 1.0 E-8
								FSH 0.5 μg/ml, somatomedin C 1 ng/ml, ovine growth hormone 6.5 μU/ml, retinoic acid 50 μg/ml
		20	100					Glutamine 3.9 E-6, nerve growth factor (NGF) 10 U/ml
			1					Hyaluronic acid 10 μg/ml, trypsin inhibitor 1 μg/ml, glucose 0.25%, glutamine 2.0 E-6
	30							NGF 5 μg/ml, Vitamin B_{12} 1.4 μg/ml, biotin 1 μg/ml, DL-α-tocopherol 10 μg/ml, retinol 5 μg/ml, cholin 1.0 E-3, carnitine 1.0 E-5, linoleic acid 1.0 E-5, lipoic acid 1.0 E-6, trace elements
	0.5		2					TRH 1 ng/ml, LRH 4ng/ml, PGE 100 ng/ml
	0.03	20	1					Thyroglobulin 100 μg/ml
				1.5 μg/cm²				

17

Table 2. continued.

Cell type	First author	Basal medium	Insulin $\mu g/ml$	Transferrin $\mu g/ml$	Selenite E-9	EGF ng/ml	FGF ng/ml	Hydrocortisone E-6
19. Hepatocytes	Castell (30)	F12	0.06					
20. Cardiomyo-cytes, rat	Kessler-Iceksen(31)	DME/F12 (1:1)	25	25				0.1
21. Myoblasts, primary, chicken	Dollen-meier (32)	MCDB 201	5	10			300	10
22. Lens epithelial cells, bovine	Plovët (33)	199/F12 (1:1)	5	5		10		
23. Erythropoietic cells	Iscove (34)	IMDM		300				
24. Erythropoietic burst cells, murine	Stewart (35)	HMEM (E) or α-MEM		300				
25. T-lymphocytes	Iscove (34)	IMDM		+				
26. B-lymphocytes	Iscove (34)	IMDM		+				
27. Adipocytes, murine (from macrophages)	Flesh (36)	DME	5	5	10			
28. Mononuclear cells, human	Schmitt (37)	RPMI 1640						
Transformed (tumour) cells								
29. Hybridoma cells	Kawamoto (38)	RPMI/ DME/F12 (2:1:1)	10	10	1			
30. Hybridoma/ myeloma lines	Cleveland (16)	IMDM/F12 (1:1)						
31. Hybridoma/ myeloma lines	Fazekas St.Groth (39)	IMDM		33				
32. T-lymphoid cell lines, human	Uitten-bogaart (40)	IMDM		1				
33. Lung adeno-carcinoma, human	Masui (41)	DME/F12 (1:1)	10	10	25	2		0.05
34. Lung small cell carcinoma, human	Carney (42)	RPMI 1640	5	100	30			0.01
35. Lung epider-moid carcin-oma, human	Masui (41)	DME/F12 (1:1)	5		25			
36. Prostatic car-cinoma line, human (PC3)	Kaighn (43)	DME/F12 (1:1)	2	2	1	25		0.05
37. Neuroblastoma mouse (C-1300)	Agy (44)	MCDB 411	1					

Dexameth-asone E-6	T₃ E-9	Proges-terone E-9	Putres-cine E-4	Fibro-nectin µg/ml	BSA mg/ml	Ethanol-amine	Phospho-ethanol-amine	Additional components and comments
0.01				+	0.2%			Glycagon 1.0 E-6, culture dishes coated with fibronectin
						2		Hepes 1.0 E-2, fetuin 1 mg/ml
				50				Glutamine 5.0 E-3, β-Hydroxybutyrate 5.0 E-4, CoCl₂ 1.0 E-7, glycerol 4.0 E-2 biotin 1µg/ml, oleate 10µg/ml
								Eye-derived growth factor 10µg/ml
					10			Fe²⁺ 7.5 E-6, oleic acid 9.2 µg/ml, L-α-dipalmitoyl lecithin 12 µg/ml, cholesterol 12 µg/ml
					10			Erythropoietin 3 U/ml, CaCl₂ 2.5 E-4, cholesterol 7.8 µg/ml, α-thioglycerol
5.0 E-5					+			Soy bean lipid 17 µg/ml (no cholesterol)
					+			Soy bean lipid 50 µg/ml (no cholesterol)
					0.1%			Glucose 4.5 g/l, pyruvate 1 E-9, NaHCO₃ 3.7 mg/l, linoleic acid 16 E-9, L-cell supernatant 30%
								Purified human serum albumin 2.5–10 mg/ml
					+			2-Mercaptoethanol 1.0 E-5, 2-aminoethanol 1.0 E-5, human low density lipoprotein 1-2 µg/ml, oleic acid 4 µg/ml complexed to fatty acid free BSA (2:1)
		10 ng/ml						α-Thioglycerol 6 nl/ml, trace elements of MCDB 104 at half conc. trace elements of MCDB 301
					1.2			Soy bean lipids 24 µg/ml
					0.4			Glutamine 0.3 mg/ml
	0.5							
								Estradiol 1.0 E-8, Selective growth of tumor cells
	0.5							Glucagon 0.2 µg/ml
	0.5							Glucagon 0.2 µg/ml, ascorbic acid 57 E-6

19

Table 2. continued.

Cell type	First author	Basal medium	Insulin µg/ml	Transferrin µg/ml	Selenite E-9	EGF ng/ml	FGF ng/ml	Hydrocortisone E-6
38. Colon carcinoma, human	van der Bosch (45)	DME/F12 (1:1)	2	2	5	20		
39. Mammary carcinoma, human (MCF-7)	Briand (46)	DME/F12 (1:1)	0.25	25	10	100		
40. Pituitary tumor cell line (GH₃)	Bottenstein (2)	F12	5	5			1	
41. Neuroblastoma rat (B104)	Bottenstein (2)	DME/F12 (1:1)	5	5	30			
42. Melanoma cell line (M2)	Bottenstein (2)	DME/F12 (1:1)	5	5	100			
43. Ovary cell line, Chinese hamster	Hamilton (47)	MCDB 301						
44. Kidney cell line	Maciag (48)	DME/F12	5	5		10	50	

Abbreviations: BSA, bovine serum albumin; DME, Dulbecco's modification of Eagle's MEM; EGF, epidermal growth factor; FGF, fibroblast growth factor; HMEM(E), Eagle's MEM with Hank's salts; IMDM, Iscove's modification of Dulbecco's medium: LRH, luteinising hormone-releasing hormone.

(ii)   reduced risk of contamination (virus, fungi, mycoplasma);

(iii)   improved economy (particularly regarding foetal calf serum);

(iv)   easier purification of culture products;

(v)   less protein interference in bioassays;

(vi)   avoidance of serum cytotoxicity;

(vii)   prevention of fibroblast overgrowth in primary cultures;

(viii) selective culture of differentiated and functional cell types from heterogeneous populations of primary cultures without preferential stimulation of variant subpopulations.

2.3 Distinct Disadvantages of using Serum in Cell Culture

As a corollary to these advantages it should be noted that there are some pronounced disadvantages of using serum:

(i)   For most cells, serum is not the physiological fluid which they contact in the original tissue from which they are derived except during wound-healing and blood-clotting processes, where serum promotes fibroblast growth, but suppresses epidermal keratinocyte growth (7).

(ii)   The potential cytotoxicity of serum is often overlooked. Besides selective inhibitors, bacterial toxins and lipids, serum may contain polyamine oxidase which, upon reaction with polyamines (like spermine, spermidine) secreted from highly proliferative cells, forms cytotoxic polyaminoaldehydes (12). Foetal calf and human pregnant serum, in contrast to horse serum, show a relatively high level of this enzyme. This may explain reports on the suppressive activity of these sera.

Dexameth- asone E-6	T_3 E-9	Proges- terone E-9	Putres- cine E-4	Fibro- nectin µg/ml	BSA mg/ml	Ethanol- amine	Phospho- ethanol- amine	Additional components and comments
	0.2							
								Glutamine 2.0 E-3, culture flask coated with collagen IV
	3.0 E-11							Somatomedin C 1 ng/ml, TRH 1 ng/ml, parathyroid hormone 0.5 ng/ml
		2.0 E-8	1.0					
		1						LRH 10 ng/ml, NGF 10 ng/ml, FSH 40 µg/ml
				5	1			

MSA, multiplication stimulating activity; NGF, nerve growth factor; PDGF, platelet derived growth factor; PGE, prostaglandin E; PGF, prostaglandin F; T_3, triiodothyronine; TRH, thyrotropine-releasing hormone.

(iii)   The relatively large batch-to-batch variations of sera require extensive serum screening, which is time-consuming and costly.

(iv)   Serum may contain inadequate levels of cell-specific growth factors, which have to be supplemented.

2.4 Potential Disadvantages of Serum-free Media

(i)   Although greater economy is possible with many cell lines, supplementing medium with hormones and growth factors can be as expensive as serum.

(ii)   Serum-free media are often highly specific to one cell type, so a different medium may be required for each cell line carried.

2.5 Who is Mainly Interested in Serum-free Media?

Everybody working on a usual laboratory scale with cell cultures and interested in controlling the culture conditions precisely may wish to avoid uncertainties imposed by batch variations of undefined serum components.

There are two main reasons applicable to *large scale cultures*, for example hybridomas to produce monoclonal antibodies, fibroblasts to produce interferon, lymphomas to produce lymphokines and other factors: (a) the avoidance of foreign proteins which must otherwise be removed and (b) the costs of fetal calf serum.

3. REPLACEMENT OF SERUM IN MEDIUM

3.1 General Procedures

Investigations to replace serum in cell cultures have followed three different lines:

(i) *Analytical approach*. Isolation and identification of each of the various serum factors required for survival and growth. A laborious task, which proved unsuccessful in the long run, since most hormonal factors act synergistically and are present only in minute quantities in serum.

(ii) *Synthetic approach by Sato and co-workers (4)*. Supplementation of existing basal, nutritional media by adding various combinations of growth factors performing serum functions: cell specific and non-cell specific hormones (including mitogens), binding proteins, attachment and spreading factors.

(iii) *Limiting factor approach by Ham and co-workers (2)*. New formulations of existing nutritional media by lowering the serum concentration, until cell growth is limited, and then adjusting the concentration of each component of the nutrient medium (i.e., vitamins, amino acids, hormones, etc.) until growth resumes. A detailed procedure has been described by Ham (2,7).

3.2 Considerations Prior to Medium Selection

3.2.1 *Non-transformed and Transformed Cells*

The reader wishing to set up a chemically defined, serum-free cell culture, should, as a first step, specify to which category the cells to be cultured belong:

(i)   non-transformed (normal) or transformed?

(ii)   is the culture intended for survival or growth?

(iii)   is the cell population homogeneous or heterogeneous (e.g., are epithelial cells mixed with fibroblasts)?

(iv)   is differentiation or proliferation required?

The following definitions may be helpful in this context (7), although it should be noted that the properties cited are not exclusive evidence for each category since there are gradual differences and exceptions to the rule.

Normal, non-transformed cell. Cell as in the healthy, intact organ; non-tumorigenic; euploid karyotype (predominantly diploid); senescence in culture with a finite number of doublings.

Transformed cell. Malignant properties; aneuploid karyotype; loss of density-dependent inhibition of growth in culture; loss of anchorage dependence; tumorigenic (facultative); low serum requirement; infinite life span.

Survival. Maintenance of viability, morphology, capacity to metabolise (facultative), and potentially, capacity to differentiate.

Growth. Cell proliferation or multiplication.

Usually, cells from primary explants of healthy tissues behave as non-transformed cells in the first generations following isolation, some for up to 50 generations or more. Such normal cells absolutely require defined quantities of high molecular mass proteins, hormones and growth factors even in the most advanced media. Transformed cells, however, may grow without these growth factors, provided that essential non-specific requirements are met. Indeed increasing evidence suggests that tumour cells are capable of stimulating their own proliferation by producing autocrine growth factors (13).

3.2.2 *Requirements of Non-transformed Cells*

Non-transformed cells usually have the following

(i) *Non-cell specific requirements:* These comprise among others: inorganic ions (Na^+, K^+, Ca^{2+}, Mg^{2+}, Fe^{2+}, Cl^-, HCO_3^-, PO_4^{3-}), CO_2, O_2, pH, osmolality, temperature, carbohydrates, nucleosides, essential amino acids, vitamins, inorganic trace elements (including selenite), insulin, transferrin, glucocorticoids and possibly somatomedins and insulin-like growth factors:

(ii) *Requirements for factors with some (but not absolute) cell specificity:* Examples are EGF, FGF, nerve growth factor (NGF), PDGF, hematopoietic growth factors (colony stimulating factors, interleukins).

Moreover it appears that several factors may express different functions (e.g., insulin seems to be essential for both proliferation and differentiation).

3.2.3 *Cell-substrate Adhesion and Biomatrix (6)*

Most non-transformed cell types must attach to a solid substrate in order to proliferate. The interaction of these cells with the environment determines many of their properties. Apart from haematopoietic cells only transformed cells show the capacity to multiply without attachment.

(i) *Substrates.* Suitable substrates for the adhesion of cells are glass surfaces or specially treated plastic materials, so-called 'tissue culture' plastics with modified surface charges. However, collagen has proved to be better than glass and plastic for cell growth and differentiation of some cells, particularly epithelial cells and neurones. Collagen is the main component of the extracellular matrix (ECM), to which cells attach. Cells do not bind directly to the substrates mentioned, but via adhesion factors, of which fibronectin, present in plasma and serum is the best characterised and most ubiquitous. Fibronectin binds to different substrates, and cells attach to this factor via specific fibronectin-receptors. There are other adhesion factors such as chondronectin, responsible for the adhesion of chondrocytes, and laminin, essential for the adhesion of epithelial cells. Some cell types are capable of synthesising collagen, laminin, glycosaminoglycans and glycoproteins *in vitro*, thus forming an ECM themselves. ECM can also be isolated as a biomatrix from different tissues and added to the tissue culture flasks or dishes. The cell attachment activity of fibronectin was recently found to reside in a tetrapeptide (L-arg-gly-1-asp-1-ser) as part of the attachment domain (14).

For procedures to coat tissue culture plastics with poly-D-lysine, collagen or fibronectin see Section 6.

(ii) *Influence of the substrate on the cells.* The substrate plays an important role on the maintenance of normal cellular functions. While isolated hepatocytes will only survive on plastic for one week, they may survive up to one month on collagen. On a biomatrix they will survive for more than 6 months. Cells cultured on plastic tend to dedifferentiate as reflected by their altered morphology. Epithelial cells assume the shape of fibroblastoid cells and start to synthesise fibroblast-specific proteins. The adhesion on collagen or biomatrix maintains the start of differentiation and reduces the necessity to add various growth factors or serum.

3.2.4 *Assays to Determine the Best Media*

The complex growth requirements of many mammalian cells impose several problems as to the adequacy of assay systems for the determination of optimal growth media. The reader is referred to a thorough review by Ham (7). Generally, a long-term (several days) cell multiplication assay is to be preferred *versus* a short-term assay using [³H]-thymidine uptake (for several hours) as the sole means to measure cell proliferation, since several artifacts are known that may simulate cell proliferation (15). Nevertheless [³H]thymidine methods are often used for screening, but the results should be cross-checked by a clonal growth assay or simply by cell counting.

3.2.5 *Media for Low-serum and Serum-free Cell Cultures*

A variety of defined media have been derived from Eagle's minimal essential medium (MEM) during the last three decades, including Dulbecco's enriched modification (DME), and rather complex media such as Ham's F12, CMRL 1066, RPMI 1640, McCoy's 5A and Iscove's modified Dulbecco's medium (IMDM). All are now commercially available and their compositions can be found in the commercial catalogues. For serum-free formulations often a 1:1 (v/v) mixture of DME and F12 is used to combine the qualitative richness of F12 (trace elements, more vitamins) with the higher nutrient concentration of DME. For haematopoietic cells, including hybridoma cultures, IMDM proved valuable when supplemented with transferrin, bovine serum albumin and several lipids. Other blood cells including leukaemic cells have been successfully grown in RPMI 1640. CMRL 1066 is particularly rich in nucleosides and some vitamins. The MCDB series was developed by Ham and co-workers especially to meet the different requirements of non-transformed cells.

The most frequently used media may be categorised (2) as:
(i)　media for permanent lines with some serum or protein supplementation: MEM, DME, α-MEM, McCoy's 5A, RPMI 1640;
(ii)　media for permanent lines with purified protein or hormone supplementation: F10, F12, DME;
(iii)　media for permanent lines in monolayer without protein supplementation: CMRL 1066, MCDB 411, DME;
(iv)　media for clonal growth of permanent lines without protein supplementation: F12, MCDB 301, DME, IMDM;
(v)　media for non-transformed cells: DME, IMDM, MCDB 104, 105, 202, 401 and 501.

Studies with various mammalian cells have documented the tissue- and species-specificity of their requirements for hormones and nutrients. Chondrocytes and fibroblasts require a lipid supplement, epithelial cells not, chondrocytes require FGF, keratinocytes, bronchial, mammary and prostatic epithelial cells require EGF, but not FGF, ethanolamine is needed by keratinocytes and bronchial epithelial cells, etc. (8).

Some media have been found generally useful for cells from several different species (2), (e.g., 199, DME, F12K and MCDB 202) but the following preferences have been suggested for individual species:
(i)　for human and monkey cells: 199, 5A, RPMI 1640, CMRL 1969, MCDB 104 and MCDB 202;

(ii)   for rat and rabbit cells: 5A, F12 and MCDB 104;

(iii) for non-transformed mouse cells: DME, CMRL 1415, MCDB 202 and MCDB 401;

(iv)   for chicken cells: 199, DME, F12K and MCDB 202.

3.2.6 *Selection of Hormones and Growth Factors*

It follows from the above discussion and from an inspection of *Table 2* that most serum-free media include insulin, transferrin and selenite. Insulin is added at relatively high concentrations ($1-10$ μg/ml), since it rapidly loses bioactivity in usual serum-free culture media. Insulin may also mimic the activity of related peptides due to peptide sequence homologies with insulin-like growth factors and somatomedins. Transferrin should be added Fe^{2+}-saturated. Among the trace elements selenite seems to be important for the activation of glutathione reductase, a key enzyme involved in O_2 metabolism and thus energy production. Glucocorticoids improve cloning efficiencies of glial cells, fibroblasts, chondrocytes, keratinocytes, bronchial, mammary and prostate eptihelial cells. Exocrine cells may also require their specific hormones (e.g., oestrogens, prolactin, progesterone for mammary cells). Some cells benefit from the supplementation with T_3, and EGF is often added to many different types of epithelial cells.

In general, no clear rules can be established that lead to a generally applicable serum-free medium. So, in many cases trial and error may be the only method to determine the best medium and to select the appropriate supplements. A well-supplemented starting medium for monolayer cultures may contain in DME/F12 (1:1, v/v): insulin 5 μg/ml, transferrin 5 μg/ml, selenite 30 nM (3.0 E-8), EGF 25 ng/ml, albumin (fatty acid- and cholesterol-free) 1 mg/ml, and fibronectin 1 μg/ml. Systematic omission of each component may identify the essential factor(s). The first-limiting-factor approach by Ham and co-workers (7) may eventually lead to optimised concentrations of the essential components. However, although straightforward and intellectually statisfying, this approach is rather time and effort consuming.

To help the beginner a list of media for 44 serum-free cell cultures with various supplements has been compiled (*Table 2*). To start, a medium may be selected that is most closely related to one already found to 'work' with similar cells. For, in general, media formulated for one cell type will support growth of other lines and primary cultures of independent origin if they are of the same cell type.

3.2.7 *Significance of Trace Elements*

The role of trace elements should not be underestimated. Cleveland *et al.* (16) recently reported that they could even discount the addition of insulin, transferrin, albumin and liposomes for serum-free hybridoma cell culture, when they expanded the number of trace elements, thus creating a totally protein- and peptide-free medium. Besides the trace elements (in ionic form) Fe, Cu, Mn, Si, Mo, V, Ni, Sn, Zn and Se they included Al, Ag, Ba, Br, Cd, Co, Cr, F, Ge, J, Rb and Zr. They regarded the elimination of albumin, transferrin and insulin as an important step since these substances are potential sources of artifacts in the use of monoclonal antibodies to identify cell surface antigens. Biotinylation of monoclonal antibodies for immunofluorescence detection of cell surface antigens will also biotinylate insulin and transferrin reacting with their corresponding receptors. Moreover commercial albumin and transferrin are far from being pure substances as they contain traces of immunoglobulins.

4. PRACTICAL SUGGESTIONS FOR MEDIUM SELECTION AND HANDLING
4.1 **General Rules and Cautions**

(i)   Make sure that preparation and handling of your medium does not remove or inactivate essential nutrients and growth factors.

   (a) Fe^{2+} may be oxidised to Fe^{3+}, may precipitate and will be removed during sterile filtration (7). Hence add Fe^{2+} just before medium sterilisation.

   (b)   Concentrated Ca^{2+}- and PO_4^{3-}-solutions may lead to the precipitation of $Ca_3(PO_4)_2$.

   (c)   High cysteine concentrations may inactivate insulin, so cysteine concentration should be reduced to 10 μM (1.0 E-5). Alternatively add another reducing agent (e.g., dithiothreitol or thioglycerol).

   (d)   Always prepare microemulsions or liposomes freshly. Unsaturated fatty acids may be oxidised; microemulsions and liposomes may be partly lost during sterile filtration. Check cytotoxicity of liposome stabiliser.

(ii)   Cells in protein-free media are more sensitive to stressful culture conditions than cells in serum-supplemented media (17). Due to higher loss of viability, transfer cells more frequently.

(iii)   Do not seed at too low cell densities ($< 10^4$ cells/ml), since cells at low inocula may not survive.

(iv)   Check quality of your culture conditions, such as quality of water (bacterial toxins, pyrogens); gas atmosphere (reduction of O_2 concentration may enhance cell growth).

(v)   Always use highest purity reagents as minor contaminants may become toxic in the absence of serum. Other fortuitous contaminants may be beneficial but if undetected will influence reproducibility, especially if batch variation in the reagent occurs.

(vi)   For some complex media (e.g., the MCDB series) separate stock solutions must be prepared and combined just before filtration. Unstable stock solutions (e.g., glutamine, Fe^{2+}) should be stored frozen and added immediately before use. Otherwise most sterile working strength media may be stored at 4°C for several weeks.

(vii)   Care should be taken when using concentrated stock solutions that all constituents remain in solution during sterile filtration. In addition, if several stock concentrations are to be combined they should be diluted as they are combined and not mixed as concentrates.

4.2 **Strategies for Medium Optimisation**

Non-transformed cells (a) or transformed cells (b)? If (a): follow strategy A; if (b): follow strategy B.

4.2.1 *Strategy A*

(i) Try to change from foetal to new born and calf serum, finally horse and human serum to reduce costs.

(ii) Reduce serum concentration down to about 2% through serial passages, if possible, to produce a nutritional adaptation.

(iii) Provide substrate for the cells to attach by coating the culture surface with: (a) poly-D-lysine; *or* (b) collagen; *or* (c) fibronectin (see Section 6).

(iv) Select medium from *Table 2*.
    The medium recommended for cells that are most closely related to yours may be tried first. Add growth factors and hormones already known to be required by similar cells. If such a medium is not available, try first a mixture of DME/F12 (1:1, v/v) with a hormone and growth factor cocktail of insulin, transferrin (Fe^{2+}-saturated) and selenite at final concentrations of 5 μg/ml, 5 μg/ml and 30 nM (3.0 E-8), respectively.

(v) In case of poor growth, add bovine serum albumin $1-5$ mg/ml and/or a cocktail of trace elements (commercially available or see, e.g., reference 16).

(vi) Control your cell detachment method prior to cell seeding. Crude trypsinisation may be harmful to your cells. Hence use pure crystalline trypsin, carefully check concentration, time and temperature (even down to 4°C) of enzymatic incubation using the microscope. Stop reaction with added trypsin inhibitor (as little as possible). Alternatively use Dispase (Boehringer Type I) instead of trypsin.

(vii) Try to reduce the albumin concentration while supplementing with hormones or growth factors (EGF, somatomedin, glucocorticoids, T_3 among others) and/or more trace elements.

(viii) If growth is still poor, check lipid requirements of your cells: provide microemulsions or liposomes containing soya bean lipids, cholesterol, linoleic acid, phosphoethanolamine and others.

(ix) Consult references $1-9$ for more suggestions.

4.2.2 *Strategy B*

(i) Follow steps (i) and (ii) of Strategy A if cells still show anchorage-dependent growth. If not, go straight to step (ii), below.

(ii) Select medium from *Table 2* following cell system No. 29. The medium recommended for cells that are most closely related to yours may be tried first. Add growth factors and hormones already known to be required by similar cells.

(iii) Determine optimum inoculum (cell density at seeding). Start at about 10^5 cells/ml. Since tumour cells usually produce their own growth factors, too low an inoculum may not yield sufficient growth factor concentrations.

(iv) Try medium of step (iv) of Strategy A, followed by the suggestions of step (v), (vii) and (viii).

For leukaemic cells try medium RPMI 1640 or CMRL 1066, eventually mixed with F12 (1:1, v/v).

Finally, the investigator should establish which ultimate goal is required with the cells to be cultured. Generally, cultured cells should reflect their origin, should show essential properties of the tissue they are derived from and should reveal the characteristics of differentiation and proliferation they exert *in situ*. While this goal may be reached with some cells, it is important to realise the limitations of present tissue culture methods for other cells. It is also possible that cultured cells will not proliferate and express differentiated functions simultaneously. Hence a period of propagation and amplification may be required followed by maintenance at high cell density ($>10^5$ cells/cm^2) to allow differentiation to occur.

5. COMMERCIALLY AVAILABLE MEDIA FOR OPTIMISED MAMMALIAN CELL CULTURE

Three categories of media may be distinguished that are offered by various companies for the optimisation of mammalian cell culture (*Table 3*).

(i) *Serum-additive*. Cells are grown in basal media containing some percentage of

Table 3. Commercially Available Serum-free Media and Supplements.

	Name	*Supplier*
Serum-additive	GF	Collaborative Research, BRL
	ITS	Collaborative Research
	SGF3	Scotts Laboratories
	SGF9	Scotts Laboratories
	SerXtend	NEN Research Products, Hana Media
Serum-substitute (undefined)	Nutricyte	Brooks Laboratories
	Nu-serum	Collaborative Research, BRL
	Ventrex	Ventrex Laboratories Inc.
	Serum Plus	
	CLEX	Dextran Products Ltd
Serum-substitute (defined)	Selectakit	Gibco
	HCGS	Biomedical Technologies
	Ultroser-G	LKB, IBF
Serum-free	HB101, HB102, HB103	NEN Research Products, Hana Media
	SF1	Northumbria Biologicals
	HL-1	Ventrex Laboratories, Paesel
	CEM	Scotts Laboratories
	MEM Iscove	Flow, Gibco, Sigma, Boehringer Mannheim, KC Biologicals
	KC-2000	KC Biologicals
	Neuman & Tytell	Gibco
	BM-86 (Wissler)	Boehringer Mannheim
	MCDB 102	Gibco
	MCDB 103,104	Flow Laboratories, Gibco

serum which is enriched by a serum-additive. Serum additives usually contain insulin, transferrin, selenite and growth factors (mostly EGF, FGF, PDGF). Products: GF, ITS; GF 3 and 9, SerXtend.

(ii) *Serum-substitute*. Cells are grown in basal, nutrient media (such as RPMI 1640/DME 1:1) supplemented with the defined serum-substitute in order to replace the serum. The substitutes usually contain insulin, transferrin, growth factors (mostly EGF, FGF, PDGF), hormones (mostly hydrocortisone or dexamethasone, T_3, progesterone). Products: Nutricyte, Hybridoma Cell Growth Supplement (HCGS), CLEX, Serum Plus, NU-Serum.

(iii) *Serum-free media*. Cells are unsupplemented in these media (DME/F12 1:1, v/v) which, in general, contain insulin, transferrin, selenite, partly albumin (fraction V), fetuin, testosterone, ethanolamine, several fatty acids, pyruvate, trace elements. Products: HB 101, HB 102, HL-1, DEM, MEM Iscove, KC-2000, BM-86 (Wissler).

6. PROCEDURES FOR COATING OF CULTURE SURFACE WITH BIOMATRIX

6.1 Poly-D-lysine Coating of Culture Surface

For anchorage-dependent cells coating the culture surface with a positively charged polymer such as poly-D-lysine has proved to be useful (2).

Dissolve 0.1 mg/ml ploy-D-lysine HBr (mol. wt. $3-7 \times 10^4$) (Sigma) in distilled water and sterilise (cellulose acetate filter, 0.22 μm).
(ii)   Pipette 0.5 ml into each 60×15 mm Petri dish.
(iii)   Rock the dish gently to spread the solution uniformly over the entire culture surface.
(iv)   After 5 min remove the solution completely with a Pasteur pipette.
(v)   Wash the dish with 1.5 ml of distilled water, rock gently and remove by aspiration.
(vi)   It is essential to remove all solutions completely, since free polylysine may be cytotoxic.
(vii)   The dishes can be used immediately or allowed to dry and used later.
(viii) Dishes coated under non-sterile conditions may be sterilised by u.v. irradiation.

6.2 Collagen Isolation and Coating of Culture Surface

Collagen type I can be obtained according to Strom and Michalopoulos (19) from the tendons that attach to the vertebrae of the rat tail.
(i)   Cut fresh or thawed tails longitudinally from the base to the tip.
(ii)   Pull the skin off the tail and remove vessels.
(iii)   Remove the white collagen fibres and sterilise with u.v. light.
(iv)   Stir 1 g of collagen fibres (from $3-4$ tails) into 300 ml of 0.1% acetic acid for 48 h at 4°C.
(v)   Filter the solution through $2-3$ sterile gauze layers.
(vi)   Dilute the stock solution 1:20 with bi-distilled water.
(vii)   Cover the bottom of a plastic dish with this solution and air dry (max. 37°C, $1-2$ days).
(viii) The dishes can be stored in a humid atmosphere at 4°C for several months.
    Alternatively, the stock solution may be neutralised and diluted in culture medium

whereupon the collagen will gel due to the increased pH and salt concentration. Denatured (air-dried) collagen enhances cell attachment (e.g., for endothelial cells and skeletal muscle), but native (undenatured) collagen gel may be required for correct phenotype expression.

6.3 Fibronectin Coating of Culture Surface

Fibronectin may be prepared according to Öbrink (18) from fresh human plasma, avoiding heparin and proteolytic degradation (by means of phenylmethylsulphonyl fluoride) or obtained commercially from BRL, Collaborative Research, or Gibco. Fibronectin is used at $1-5$ μg/cm^2 growth area of the culture dish.

(i)   Dissolve 20 μg of fibronectin lyophilisate gently in 1 ml of bi-distilled water.

(ii)   Cover the bottom of the dish (35 mm diameter) with solution and air-dry at 40°C.

(iii) Rinse extensively with distilled water and air-dry again.

(iv)   Coated dishes can be stored at room temperature for several months.

7. REFERENCES

1. Freshney,R.I. (1983) *Culture of Animal Cells − A Manual of Basic Technique*, Alan Liss, Inc., NY, p. 74.
2. Ham,R.G. and McKeehan,W.L. (1979) in *Methods in Enzymology,* Vol. **53**, Jacoby,W.B. and Pastan,I.H. (eds.), Academic Press, NY, p. 44.
3. Bottenstein,J., Hayashi,I., Hutchings,S., Masui,H., Mather,J., McClure,D.B., Ohasa,S., Rizzino,A., Sato,G., Serrero,G., Wolfe,R. and Wu,R. (1979), in *Methods in Enzymology, Vol. 53: Cell Culture*, Jakoby,W.B. and Pastan,I.H. (eds.), Academic Press, NY, p. 94.
4. Barnes,D. and Sato,G. (1980) *Anal. Biochem.*, **102**, 255.
5. Barnes,D. and Sato,G. (1980) *Cell*, **22**, 649.
6. Sato,G.H., Pardee,A.B. and Sirbasku,D.A. (eds.) (1982) *Growth of Cells in Hormonally Defined Media*, Cold Spring Harbor Conf. Cell Prolif. Vol. 9, Book A and B, Cold Spring Harbor Laboratory Press, NY.
7. Ham,R.G. (1981) in *Tissue Growth Factors*, Baserga,R. (ed.), Handbook of Experimental Pharmacology, Vol. 57, p. 13.
8. Ham,R.G. (1983) in *Hormonally Defined Media*, Fischer,G. and Wieser,R.J. (eds.), Springer Verlag, Berlin, Heidelberg, NY, Tokyo, p. 16.
9. Fischer,G. and Wieser,R.J. (eds.) (1983) *Hormonally Defined Media*, Springer Verlag, Berlin, Heidelberg, NY, Tokyo.
10. Bradshaw,R.A. and Rubin,J.S. (1980), *J. Supramol. Struct.*, **14**, 183. Contr. Cell Div. Dev. Part A: 193-209.
11. Iscove,N. (1982) First Europ. Conf. Serum-Free Cell Culture, Heidelberg.
12. Ali-Osman,F. and Maurer,H.R. (1983) *J. Cancer Res. Clin. Oncol.*, **106**, 17.
13. Bürk,R.R. (1983) in *Hormonally Defined Media*, Fischer,G. and Wieser,R.J. (eds.), Springer Verlag, Heidelberg, p. 57.
14. Pierschbacher,M.D. and Ruoslahti,E. (1984) *Nature*, **309**, 30.
15. Maurer,H.R. (1981) *Cell Tissue Kinetics*, **14**, 111.
16. Cleveland,W.L., Wood,I. and Erlanger,B.F. (1983) *J. Immunol. Methods*, **56**, 221.
17. McHugh,Y.E., Walthall,B.J. and Steimer,K.S. (1983) *Biotechniques June/July*, p. 72.
18. Öbrink,B. (1982) in *Methods in Enzymology*, Vol. **82**, Cunningham,L.W. and Fredericksen,D.W. (eds.), Academic Press, NY, p. 513.
19. Strom,S.C. and Michalopoulos,G. (1982) in *Methods Enzymology*, Vol. **82**, Cunningham,L.W. and Fredericksen,D.W. (eds.), Academic Press, NY, p. 544.
20. Clark,J. (1983) in *Hormonally Defined Media*, Fischer,G. and Weiser,R.J. (eds.), Springer Verlag, Berlin, Heidelberg, NY, Tokyo, p. 6.
21. McKeehan,W.L., McKeehan,K.A., Hammond,S.L. and Ham,R.C. (1977) *In vitro*, **13**, 399.
22. Jennings,S.D. and Ham,R.G. (1981) *In Vitro*, **17**, 238 (Abtract No. 155).
23. Adolphe,M., Ronot,Z., Froger,B., Corvol,M.T. and Forest,N. (1983) in *Hormonally Defined Media*,

Fischer,G. and Wieser,R.J. (eds.), Springer-Verlag, Berlin, Heidelberger, NY, Tokyo, p. 377.
24. Chaproniere-Rickenburg,D.M. and Weber, M.M. (1982) *Cold Spring Harbor Conf. Cell Prolif.*, **9**, 1109.
25. Bunge,R.P., Carey,D.J., Higgins,D., Eldridge,C. and Roufa,D. (1983) in *Hormonally Defined Media*, Fischer,G. and Wieser,R.J. (eds.), Springer Verlag, Berlin, Heidelberg, NY, Tokyo, p. 178.
26. Fischer,G. (1983) in *Hormonally Defined Media*, Fischer,G. and Wieser,R.J. (eds.), Springer Verlag, Berlin, Heidelberg, NY, Tokyo, p. 189.
27. Honegger,P. and Güntert,B. (1983) in *Hormonally Defined Media*, Fischer,G. and Wieser,R.J. (eds.), Springer Verlag, Berlin, Heidelberg, NY, Tokyo, p. 203.
28. Lang,E., Lang,K., Krause,U., Racke,K., Nitzgen,B. and Brunner,G. (1983) in *Hormonally Defined Media*, Fischer,G. and Wieser,R.J. (eds.), Springer Verlag, Berlin, Heidelberg, NY, Tokyo, p. 250.
20. Orly,J., Weinberger-Ohana,P. and Farkash,Y. (1983) in *Hormonally Defined Media*, Fischer,G. and Wieser,R.J. (eds.), Springer Verlag, Berlin, Heidelberg, NY, Tokyo, p. 274.
30. Castell,J.V., Gómez-Lechon,M.D., Coloma,J. and Lopez,P. (1983) in *Hormonally Defined Media*, Fischer,G. and Wieser,R.J. (eds.), Springer Verlag, Berlin, Heidelberg, NY, Tokyo, p. 333.
31. Kessler-Icekson,G., Wassermann,L., Yoles,E. and Sampson,S.R. (1983) in *Hormonally Defined Media*, Fischer,G. and Wieser,R.J. (eds.), Springer Verlag, Berlin, Heidelberg, NY, Tokyo, p. 383.
32. Dollenmeier,P. and Eppenberger,H.M. (1983) in *Hormonally Defined Media*, Fischer,G. and Wieser,R.J. (eds.), Springer Verlag, Berlin, Heidelberg, NY, Tokyo, p. 358.
33. Plouët,J., Olivié,M., Courty,J., Courtois,Y. and Barritault,D. (1983) in *Hormonally Defined Media*, Fischer,G. and Wieser,R.J. (eds.), Springer Verlag, Berlin, Heidelberg, NY, Tokyo, p. 123.
34. Iscove,N.N. and Melchers,F. (1978) *J. Exp. Med.*, **147**, 923.
35. Stewart,S., Bi-de Zhu and Axelrad,A. (1984) *Exp. Hematol.*, **12**, 309.
36. Flesch,I., Ketelsen,U.P. and Ferber, E. (1983) in *Hormonally Defined Media*, Fischer,G. and Wieser,R.J. (eds.), Springer Verlag, Berlin, Heidelberg, NY, Tokyo, p. 304.
37. Schmitt,S. and Schenkein,H.A. (1983) *J. Immunol. Methods*, **63**, 337.
38. Kawamoto,T., Sato,J.D., Le,A., McClure,D.B. and Sato,G.H. (1983) in *Hormonally Defined Media*, Fischer,G. and Wieser,R.J. (eds.), Springer Verlag, Berlin, Heidelberg, NY, Tokyo, p. 310.
39. Fazekas de St.Groth,S.F. (1983) *J. Immunol. Methods*, **57**, 121.
40. Uittenbogaart,C.H., Cantor,Y. and Fahey,J.L. (1983) *In Vitro*, **19**, 67.
41. Masui,H., Miyazaki,K. and Sato,G.H. (1983) in *Hormonally Defined Media*, Fischer,G. and Wieser,R.J. (eds.), Springer Verlag, Berlin, Heidelberg, NY, Tokyo, p. 430.
42. Carney,D.N., Bunn,P.A.,Jr., Gazdar,A.F., Pagan,J.A. and Minna,J.D. (1981) *Proc. Natl. Acad. Sci. USA*, **78**, 3185.
43. Kaighn,M.E. (1983) in *Hormonally Defined Media*, Fischer,G. and Wieser,R.J. (eds.), Springer Verlag, Berlin, Heidelberg, NY, Tokyo, p. 418.
44. Agy,P.C., Shipley,G.D. and Ham,R.G. (1981) *In Vitro*, **17**, 671.
45. van der Bosch,J. (1983) in *Hormonally Defined Media*, Fischer,G. and Wieser,R.J. (eds.), Springer Verlag, Berlin, Heidelberg, NY, Tokyo, p. 412.
46. Briand,P. and Lykkesfeldt,A.E. (1983) in *Hormonally Defined Media*, Fischer,G. and Wieser,R.J. (eds.), Springer Verlag, Berlin, Heidelberg, NY, Tokyo, p. 436.
47. Hamilton,W.G. and Ham,R.G. (1977) *In Vitro*, **13**, 537.
48. Maciag,T., Kelley,B., Cerundolo,J., Ilsley,S., Kelly,P.R., Gaudreau,J. and Forand,R. (1980) *Cell Biol. Int. Rep.* **4**, p. 43.

CHAPTER 3

Scaling-up of Animal Cell Cultures

B. GRIFFITHS

1. INTRODUCTION

Small cultures of cells in, for example, flasks of up to 1 l volume (175 cm² surface area), are the best means for establishing new cell lines in culture, for studying cell morphology, and for comparing the effects of different agents, or different concentrations of an agent, on growth and metabolism. However, there are many applications in which large numbers of cells are required, for example, extraction of a cellular constituent (10^9 cells will provide 7 mg DNA); to produce viruses for vaccine production (typically 5×10^{10} cells per batch) or other cell products (interferon, plasminogen activator, various hormones, enzymes and antibodies); and to produce inocula for even larger cultures. If large quantities of cells are needed then setting up multiples of small cultures is tedious, labour-intensive and expensive. One obvious route would be to scale-up the size of the culture vessel, to develop a 'unit process' system. Unfortunately, it is not enough to scale-up proportionally a particular culture vessel. As the size increases parameters change and modifications to the system have to be made. One of the aims of this chapter is to point out what these changes are, at what scale they must be implemented, and possible solutions to the problems they create. This theme has to be applied to cells that grow in suspension and cells that will only grow when attached to a substrate (anchorage-dependent cells).

Suspension culture offers the easiest means of scale-up because a 1 l vessel is conceptually very similar to a 1000 l vessel. The changes concern the degree of environmental control and the means of maintaining the correct physiological conditions for cell growth, rather than significantly altering vessel design. Monolayer systems (for anchorage-dependent cells) are very difficult to scale-up in a unit system, as opposed to a multiple process, and consequently a wide range of diverse systems have evolved. The aim has been to increase the surface area available to the cells in relation to both the medium and the total culture volume.

Before describing monolayer and suspension culture techniques and equipment in detail some of the more generalised aspects of cell culture are discussed, many of which are common to both systems.

2. GENERAL METHODS AND CULTURE PARAMETERS

Familiarity with certain biological concepts and methods is essential when undertaking scale-up of a culture system. In small scale cultures there is often some leeway for error. If the culture fails it is a nuisance, but not necessarily a disaster in terms of time and costs. As the size of the culture increases it represents an ever-increasing investment of resources. Culture failure is then more serious but at the same time the system increasingly demands that all conditions are met more critically. This section describes the basic ingredients of all cell culture systems which have to be attended to as the culture size gets larger and more diverse in design and operation.

2.1 Cell Quantification

Measuring total cell numbers (by haemocytometer counts of whole cells or stained nuclei) and total cell mass (by determining protein or dry weight) is easily achieved. It is far more difficult to get a reliable measure of cell viability because the methods employed either stress the cells or use a specific, and not necessarily typical, parameter of cell physiology (see Chapter 8). An additional difficulty is that in many culture systems the cells cannot be sampled (most anchorage-dependent cultures), or visually examined, and an indirect measurement has to be made.

2.1.1 Cell Viability

(i) *Dye exclusion*. The dye exclusion test is based on the concept that viable cells do not take up certain dyes, whereas dead cells are permeable to these dyes. Trypan blue (0.4%) is the most commonly used dye, but has the disadvantage of staining soluble protein. In the presence of serum, therefore, erythrosine B (0.4%) is often preferred (see Chapter 4). Cells are enumerated in the standard manner using a haemocytometer. Some caution should be used when interpreting results as the uptake of the dye is pH- and concentration-dependent, and there are situations in which misleading results can be obtained. Two relevant examples are membrane leakiness caused by recent trypsinisation and freezing and thawing in the presence of dimethylsulphoxide. Further details on measuring cell viability are given in Chapters 4 and 8.

2.1.2 Indirect Measurements

Indirect measurements of viability are based on metabolic activity. The most commonly used parameter is glucose utilisation but oxygen utilisation, lactic or pyruvic acid production, or carbon dioxide production can also be used, as can the expression of a product, such as an enzyme. When cells are growing logarithmically, there is a very close correlation between nutrient utilisation and cell numbers. However, during other growth phases high utilisation rates, caused by maintenance rather than growth, can give misleading results. The measurements obtained can be expressed as a growth yield (Y) or specific utilisation/respiration rate (Q):

$$\text{Growth yield (Y)} = \frac{\text{Biomass concentration (DX)}}{\text{Substrate concentration (DS)}} \qquad \text{Equation 1}$$

$$\text{Specific utilisation/respiration rate (QA)} = \frac{\text{Substrate concentration (DS)}}{\text{Time (DT)} \times \text{cell mass/ numbers (X)}} \qquad \text{Equation 2}$$

Typical values of growth yields for glucose (10^6 cells/g) are 385 (MRC-5), 620 (Vero) and 500 (BHK).

2.2 **Equipment and Reagents**

2.2.1 *Culture Vessel and Growth Surfaces*

The standard non-disposable material for growth of animal cells is glass, although this is replaced by stainless steel in larger cultures. It is preferable to use borosilicate glass (e.g., Pyrex) because it is less alkaline than soda glass. The primary requirement is that glassware is thoroughly cleaned and rinsed at least three times in deionised water. Cells attach readily to glass, but attachment may be augmented by various surface treatments (see Section 3.1). In suspension culture, cell attachment has to be discouraged, and this is achieved by treatment of the culture vessel with a proprietary silicone preparation. Examples are Dow Corning 1107 (which has to be baked on) or dimethyldichlorosilane (Repelcote, Hopkins and Williams) which requires thorough washing of the vessel in distilled water to remove the trichloroethane solvent.

2.2.2 *Tubing*

As culture systems become larger and more sophisticated, so the need to provide connections between the various units in the system becomes greater. These connections should be a combination of silicone and stainless steel tubing. Silicone tubing is very pervious to gases, which can cause problems when transferring media due to loss of dissolved CO_2. It is also liable to rapid wear when used in a peristaltic pump. Thick walled tubing with additional strengthening (sleeve) should be used. When connecting silicone tubing to stainless steel, secure with plastic ties to prevent blow off during autoclaving or if back pressure builds up in the system during use. Also, use silicone tubing tips on the end of stainless tubes to prevent scratching the culture vessel, or causing mechanical damage to the cells.

2.3.3 *Connectors*

It is often necessary to connect alternative vessels to the culture system during the culture period, or even to assemble after autoclaving in separate units. This has to be done aseptically and the investment in properly machined screw-together connectors is costly but very worthwhile (e.g., the range marketed by LH Fermentation).

2.2.4 *Sampling Medium or Cells in Suspension*

Safe removal of samples of the culture at frequent intervals is essential. An entry with a vaccine stopper through which a hypodermic syringe can be inserted provides a simple solution, but is only suitable for small cultures. Repeated piercing of the vaccine stopper can lead to a loss of culture integrity. The use of a sampling device, such as the one developed at AVRI (Pirbright, Surrey, UK) (*Figure 1*), is a good investment as it automatically enables the line to be cleared of static medium containing dead cells and thus avoids the necessity of taking small initial samples which are then discarded.

2.2.5 *Filters*

Air filters are required for the entry and exit of gases. Even if continuous gassing is not used, one filter entry is usually needed to equilibrate pressure and for forced input

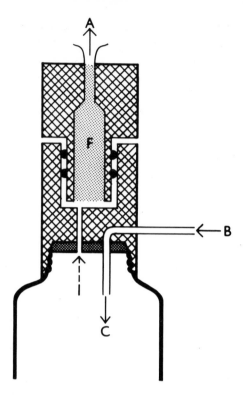

Figure 1. Culture sampling device developed by AVRI, Pirbright and manufactured by LH Fermentation. Draw out air from A (e.g., with a syringe) and medium will be drawn from the culture via B into the bijou bottle C.F. is a cotton wool filter.

or withdrawal of medium. The filters should have a 0.22 μm rating and be non-wettable. Suitable examples are the Microflow 50 and Pall Ultipor.

2.2.6 *Non-nutritional Medium Supplements*

(i) Sodium carboxymethyl cellulose (15 − 20 c.p.s.) is often added to media (at 0.1 % concentration) to help minimise mechanical damage to cells caused by the shear force generated by the stirrer impeller. This compound is more soluble than methylcellulose, and is one of the compounds that has a higher solubility at 4°C than at 37°C.

(ii) Pluronic F-68 (polyglycol) (BASF, Wyandot Corp. Inc., USA) is often added to media (at 0.1 % concentration) to reduce the amount of foaming that occurs in stirred and/or aerated cultures, especially when serum is present. It is also helpful in reducing cell attachment to glass by suppressing the action of serum in the attachment process.

2.3 **Practical considerations**

(i) Always pre-warm the medium to the operating temperature (usually 37°C) and stabilise the pH before adding the cells. Shifts in pH during the initial stages of a culture create many problems including a long lag phase and reduced yield.

(ii)   Avoid using stationary phase cells as an inoculum since this will mean a long lag phase, or no growth at all. Ideally late logarithmic phase cells should be used.

(iii) Always inoculate at a high enough cell density. There is no set rule as to the minimum inoculum level below which cells will not grow, as this varies bet-. ween cell lines and the complexity of the medium being used. A guide is bet- ween 5×10^4 and 2×10^5 cells/ml, or 5×10^3 and 2×10^4 cells/cm².

(iv)   Find empirically the optimum stirring rate for a given culture vessel and cell line. This could vary between 100 and 500 r.p.m. for suspension cells, but is usually within the range $200-350$ r.p.m., and between 20 and 100 r.p.m. for microcarrier cultures.

(v)   The productivity of the system depends upon the quality and quantity of the medium and, for anchorage-dependent cells, the surface area for cell growth. A unit volume of medium is only capable of giving a finite yield of cells. Fac- tors which affect the yield are: pH; oxygen limitation; accumulation of toxic metabolites; nutrient limitation; spatial restrictions; and mechanical stress.

As soon as one of these factors comes into effect, the culture is finished and the re- maining resources of the system are wasted. The aim is, therefore, to delay the onset of any one factor until the accumulated effect causes cessation of growth at which point the system has been optimally utilised. Simple examples of achieving this are: a better buffering system (e.g., Hepes instead of bicarbonate); continuous gassing; generous headspace, use of complete (rather than minimal) media with nutrient sparing sup- plements such as lactalbumin hydrolysate, peptone, etc.; perfusion loops, especially through filter fibres, or dialysis tubing, for detoxification; and attention to culture design.

2.4 Growth Kinetics

The standard format of a culture cycle beginning with a lag phase, proceeding through the logarithmic phase to a stationary phase, and finally to the decline and death of cells (*Figure 2*), is well documented (see Chapter 1). Although cell growth usually refers

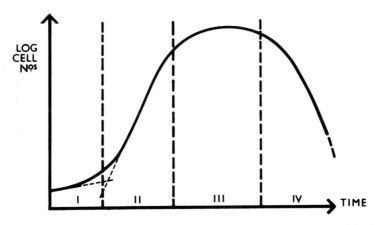

Figure 2. Culture growth phases. I, lag; II, logarithmic; III, stationary; IV, decline and death.

to increase in cell numbers, increase in cell mass can occur without any replication. The difference in mean cell mass between different cell populations is considerable, as would be expected, but so is the variation within the same population.

Growth (increase in cell numbers or mass) can be defined in the following terms.

(a) Specific growth rate (μ) [i.e., the rate of growth per unit amount (weight/numbers) of biomass]:

$$\mu = (1/x) \, (dx/dt) h^{-1} \qquad \text{Equation 3}$$

whre dx = increase in cell mass; dt = time interval; and x = cell mass.
If the growth rate is constant (e.g., during logarithmic growth) then:

$$\ln_x = \ln x_o + \mu t \qquad \text{Equation 4}$$

where x_o = biomass at time t_o.

(b) Doubling time (td) (i.e., the time for a population to double in number/mass):

$$td = \frac{\ln 2}{\mu} = \frac{0.693}{\mu} \qquad \text{Equation 5}$$

(c) Degree of multiplication (n) òr number of doublings (i.e., the number of times the inoculum has replicated):

$$n = 3.32 \log (x/x_o) \qquad \text{Equation 6}$$

2.5 Medium and Nutrients

A given concentration of nutrients can only support a certain number of cells. Alternative nutrients can often be found by a cell when one becomes exhausted, but this is bad practice because the growth rate is always reduced (e.g., while alternative enzymes are being induced). If a minimal medium, such as Eagle's basal medium (BME) or minimum essential medium (MEM) is used with serum as the only supplement, then this problem is going to be met sooner than in cultures using complete media (e.g., 199), or in media supplemented with lactalbumin hydrolysate, peptone or bovine serum albumin (which provides many of the fatty acids). Nutrients likely to be exhausted first are glutamine, partly because it spontaneously cyclises to pyrrolidone carboxylic acid and is enzymically converted (by serum and cellular enzymes) to glutamic acid, leucine and isoleucine. Human diploid cells are almost unique in utilising cystine heavily. The many reports of arginine being a limiting factor are misleading, as often these cells were contaminated with mycoplasma. A point to remember is that nutrients become growth-limiting before they become exhausted. As the concentration of amino acids falls, the cell finds it increasingly difficult to maintain sufficient intracellular pool levels. This is exaggerated in monolayer cultures because as cells become more tightly packed together, the surface area available for nutrient uptake becomes smaller (1).

Glucose is often another limiting factor as it is destructively utilised by cells and, rather than adding high concentrations at the beginning, it is more beneficial to supplement after 2 – 3 days. In order to maintain a culture some additional feeding often has to be carried out either by complete, or partial, media changes or by perfusion. The data in *Figure 3* compare the growth stimulation to MRC-5 cells as a result of adding equal quantitites of media by continuous perfusion (over 24 h) and media changes. The efficiency of medium changes is probably due to the high extracellular concentration

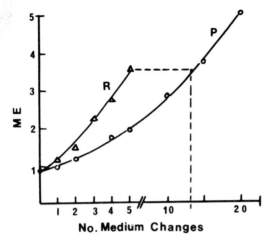

Figure 3. Comparison of MRC-5 growth using continuous perfusion (P) and complete medium changes (R). The same total volumes of medium are used per day for both systems. Growth is given as monolayer equivalents (ME) to demonstrate multilayering of MRC-5 cells.

of nutrients this provides, thus stimulating a further replicative cycle. The warning these data provide is that if perfusion is used, ensure that the perfusion rate is high enough to optimise the growth conditions.

Many cell types are either totally dependent upon, or can only perform optimally, when certain growth factors are present (see Chapter 2).

Cell aggregation is often a problem in suspension cultures. Media lacking calcium and magnesium ions have been designed specifically for suspension cells because of the role of these ions in attachment (see Section 3.1). This problem has also been overcome by including very low levels of trypsin in the medium (2 μg/ml).

2.6 pH

Ideally pH should be near 7.4 at the initiation of a culture and not fall below a value of 7.0 during the culture, although many hybridoma lines appear to prefer a pH of around 7.0. A pH below 6.8 is usually inhibitory to cell growth. Factors affecting the pH stability of the medium are: buffer capacity and type; headspace; and glucose concentration.

The normal buffer system in tissue culture media is the CO_2-bicarbonate system analogous to that in blood. This is a weak buffer system in that it has a pKa well below the physiological optimum. It also requires the addition of CO_2 to the headspace above the medium to prevent loss of CO_2 and increase in hydroxyl ions. The buffering capacity of the medium is increased by the phosphates present in the balanced salt solution (BSS). Media intended to equilibrate with 5% CO_2 usually contain Earle's BSS (25 mM $NaHCO_3$) but an alternative is Hanks' BSS (4 mM $NaHCO_3$) for equilibration with air. Improved buffering and pH stability in media is possible by using a zwitterionic buffer (2), such as Hepes (between $10-20$ mM), either in addition to, or instead of, bicarbonate (include bicarbonate at 0.5 mM). Alternative buffer systems are provided by using specialist media such as Leibovitz's L-15 (3). This medium utilises the buffering capacity of free amino acids, substitutes galactose and pyruvate for glucose, and omits sodium bicarbonate. It is suitable for open cultures.

The headspace volume in a closed culture is important because in the initial stages of the culture 5% CO_2 is needed to maintain a stable pH in the medium but, as the cells grow and generate CO_2, then a build up of CO_2 in the headspace will prevent CO_2 diffusing out of the medium. The result is an increase in weakly dissociated $NaHCO_3$ producing an excess of H^+ ions in the medium and a fall in pH. Thus, a large headspace is required in closed cultures, typically 10-fold greater than the medium volume (this volume is also needed to supply adequate oxygen). This generous headspace is not possible as cultures are scaled-up in size and an open system with a continuous flow of air, supplied through one filter and extracted through another, is required.

The metabolism of glucose by cells results in the accumulation of pyruvic and lactic acids. Glucose is metabolised at a far greater rate than it is needed. Thus, glucose should ideally never be included in media at concentrations above 2 g/l and it is better to supplement during the culture than to increase the initial concentration. An alternative is to substitute glucose with galactose or fructose as this significantly reduces the formation of lactic acid, but it usually results in a slower growth rate. These precautions delay the onset of a non-physiological pH and are sufficient for small cultures. As scale-up increases, headspace volume and culture surface area in relation to the medium volume decrease. Also, many systems are developed in order to increase the surface area for cell attachment and cell density per unit volume. Thus, pH problems occur far earlier in the culture cycle as CO_2 cannot escape as readily, and more cells means higher production of lactic acid and CO_2. The answer is to carry out frequent medium changes, or have a pH control system.

The basis of a pH control system is summarised in *Figure 4*. An autoclavable pH probe (available from Pye Ingold, Russell) feeds a signal to the pH controller and this is converted to give a digital or analogue display of the pH. This is a pH monitor system. Control of pH requires the defining of high and low pH values beyond which the pH should not go. These two set points on the pH scale turn on a relay to activate a pump, or solenoid valve, which will allow additions to be made to the culture to bring it back to within the allowable units. The more common agents that are used are shown in *Figure 4*. It is rare to have the pH rise above the set point once the medium has equili-

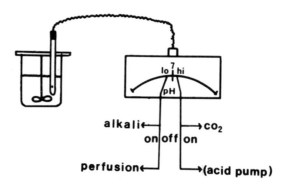

Figure 4. Diagram of pH control of cell cultures. Lo, low set point; Hi, high set point on pH scale.

brated with the headspace, so the addition of acid can be disregarded. If an alkali is
added then sodium bicarbonate (5.5%) is recommended. Sodium hydroxide (0.2 M)
can only be used with a fast stirring rate, which dilutes the alkali before localised concentrations can damage the cells, or if a perfusion loop is installed. Normally liquid
delivery pumps are supplied as part of the pH controller. Gas supply is controlled by
a solenoid valve. A recommended practice is to supply the culture with 5% CO_2 through
a solenoid valve and 95% air directly. Above the set point CO_2 will be mixed with
the air but below this point only air will be delivered to the culture. This in itself is
a controlling factor in that it helps remove CO_2, as well as meeting the oxygen requirements of the cultures.

2.7 Oxygen

The scale-up of animal cell cultures is very dependent upon the ability to supply sufficient oxygen without causing cellular damage. Oxygen is only sparingly soluble in culture
media (7.6 μg/ml) and a survey of reported oxygen utilisation rates by cells (4) reveals
a mean value of 6 μg/10^6 cells/h. A typical culture of 2×10^6 cells/ml would, therefore,
deplete the oxygen content of the medium (7.6 μg/ml) in under 1 h. It is necessary
to supply oxygen to the medium throughout the life of the culture and the ability to
do this adequately depends upon the oxygen transfer rate (OTR) of the system:

$$OTR = Kla\ (C* - C) \qquad \text{Equation 7}$$

where OTR = amount of O_2 transferred/unit volume/unit time; Kl = oxygen transfer
coefficient; a = area of the interface across which oxygen transfer occurs [as this can
only be measured in stationary and surface aerated cultures, the value Kla is used; Kla
= mass transfer coefficient (vol/h)]; C* = concentration of dissolved oxygen when
medium is saturated; and C = actual concentration of oxygen at any given time.

The Kla (OTR/C* when C = 0) is given as h^{-1} and is thus a measure of the time
taken to completely oxygenate a given culture vessel under a particular set of conditions.

A culture can be aerated by one, or a combination, of the following methods: surface aeration; sparging; membrane diffusion; medium perfusion; increasing the partial
pressure of O_2; and increasing the atmospheric pressure.

2.7.1 *Surface Aeration − Static Cultures*

In closed systems, such as a sealed flask, the important factors are the amount of oxygen in the system and the availability of this oxygen to the cells growing under 3−6 mm
of medium. Normally a headspace to medium volume ratio of 10:1 is used in order
to provide sufficient oxygen. Thus a 1 l flask (e.g., Roux bottle) with 900 cc of air
and 100 cc medium will initially contain 0.2776 g oxygen (*Table 1*). This amount will
support 10^8 cells for 450 h and is thus clearly adequate. The second factor is whether
this oxygen can be made available to the cells. The transfer rate of oxygen from the
gas phase into a liquid phase has been calculated at about 17 μg/cm²/h (4). Again, this
is well in excess of that required by cells in a 1 l flask. However, if the surface is
assumed to be saturated with dissolved oxygen and the concentration at the cell sheet
is almost zero then, applying Fick's law of diffusion, the rate at which oxygen can
diffuse to the cells is about 1.5 μg/cm²/h. At this rate there is only sufficient oxygen
to support about 50×10^6 cells in a 1 l flask, a cell density which in practice many

Table 1. Oxygen Concentrations on the Gas and Liquid Phases of a Roux Bottle Culture.

Oxygen in 900 ml air:-
$900 \times 0.21 \times 32/22400 = 0.27$ g
Oxygen in 100 ml medium:-
$100 \times 7.6 \times 10^{-4} = 0.0076$ g

Notes:	
0.21	= proportion of oxygen in air
32	= molecular weight of oxygen
22400	= gram-molecular-volume
7.6×10^{-4}	= solubility of oxygen in water at 37°C when equilibrated with air

tissue culturists take as the norm. These calculations show the importance of maintaining a large headspace volume otherwise oxygen limitation could become one of the growth-limiting factors in static closed cultures. A fuller explanation of these calculations, including Fick's law of diffusion, can be found in a review by Spier and Griffiths (4).

2.7.2 Surface Aeration — Stirred Cultures

Oxygen transfer into a stirred culture is dependent upon the stirring rate and the geometry of the impeller. The Kla has to be found by experimentation (4) but in general terms aeration is not significantly increased above surface aeration rates until the r.p.m. exceeds 100. As the oxygen entering the system is largely a function of the surface area of the culture, it is important in scale-up not to exceed a 2:1 ratio of medium depth to diameter, unless a more efficient aeration system is employed.

2.7.3 Sparging

This is the bubbling of gas through the culture and is a very efficient means of effecting oxygen transfer (as proven in bacterial fermentation). However, it is damaging to animal cells due to the effect of the high surface energy of the bubble on the cell membrane. This damaging effect can be minimised by using large air bubbles (which have lower surface energies than small bubbles), by using a very low gassing rate (e.g., 5 cc/l/min) or by adding pluronic F-68. A specialised form of sparging is described in Section 4.7 (airlift fermenter) which has also been used successfully in large unit process monolayer cultures (e.g., multiple plate propagators). When sparging is used, efficiency of oxygenation is increased by using a large height to diameter ratio culture vessel. This creates a higher pressure at the base of the reactor which increases oxygen solubility.

2.7.4 Membrane Diffusion

Silicone tubing is very pervious to gases and if long lengths of thin walled tubing can be arranged in the culture vessel then sufficient diffusion of oxygen into the culture can be obtained. However, large lengths are required (e.g., 30 m of 2.5 cm tubing for a 1000 l culture). This method is expensive and inconvenient to use and has the inherent problem that scale-up is mainly two-dimensional while that of the culture is three-dimensional.

2.7.5 *Medium Perfusion*

A closed loop perfusion system continuously (or on demand) takes medium from the culture and passes it through an oxygenation chamber before it is returned to the culture. This method has many advantages if the medium can be conveniently separated from the cells for perfusion through the loop. The medium in the chamber can be vigorously sparged to ensure oxygen saturation and other additions, such as sodium hydroxide for pH control which would damage the cells if put directly into the culture, can be made. This method is used in glass bead systems (Section 3.6.1) and has proved particularly effective in microcarrier systems (Section 3.6.3).

2.7.6 *Environmental Supply*

The dissolved oxygen concentration can be increased by increasing the headspace pO_2 (from atmospheric 21% to any value using oxygen and nitrogen mixtures) and by raising the pressure of the culture by 1 atmosphere (which increases the solubility of oxygen and its diffusion rate). These methods should only be employed when the culture is well advanced otherwise oxygen toxicity effects could occur. Finally, the geometry of the stirrer blade also affects the oxygen transfer rate.

2.7.7 *Scale-up*

Oxygen limitation is usually the first factor to be overcome in culture scale-up. This becomes a problem in conventional stirred cultures at volumes above 10 l. However, with the current use of high density cultures maintained by perfusion, oxygen limitation can occur in a 2 l culture. In order to predict the point at which this can occur, a computer model similar to that described by Spier and Griffiths (4) can be used. This model requires the user to input data on cell growth characteristics (type, inoculum, lag phase and mean generation time) and culture characteristics (medium and headspace volume, surface area, gas composition and pressure). The programme allows choice of oxygenation method (listed in Section 2.7) in both stirred and stationary cultures. The oxygen balance is calculated at 8-h intervals and values for cell yield, oxygen supplied and utilised are given. Oxygenation parameters such as r.p.m. and air flow/sparging rates can be varied.

2.7.8 *Redox Potential*

The oxidation-reduction (ORP), or redox potential, is a measure of the charge of the medium and is thus affected by the proportion of oxidative and reducing chemicals, the oxygen concentration and the pH. When fresh medium is prepared and placed in the culture vessel it takes time for the redox potential to equilibrate, a phenomenon known as poising. The optimum level for growth of many cell lines is + 75 mV, which corresponds to a dissolved pO_2 of 8 − 10%. Some investigators find it beneficial to control the oxygen supply to the culture by means of a redox, rather than an oxygen, electrode. Alternatively, if the redox potential is monitored by means of a redox electrode and pH meter (with mV display), then an indication of how cell growth is progressing can be obtained (5). This is because the redox value falls during logarithmic growth and reaches a minimum value approximately 24 h before the onset of the stationary

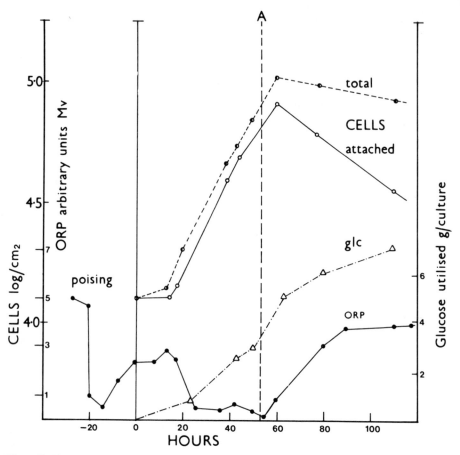

Figure 5. Changes in ORP correlated to cell growth and glucose utilisation (6). The minimum ORP value is reached 24 h before the end of logarithmic growth (A).

phase (*Figure 5*). This provides a useful guide to cell growth in cultures where cell sampling is not possible. It is also useful to be able to predict the end of the logarithmic growth phase so that medium changes, addition of virus or product promoters can be given at the optimum time. The effect of redox potential on cell cultures has been reviewed by Griffiths (6).

2.8 Types of Culture Systems

2.8.1 *Batch*

This is the usual type of culture in which cells are inoculated into a fixed volume of medium. As they grow, nutrients are consumed and metabolites accumulate. The environment, therefore, is continually changing and this, in turn, enforces changes to cell metabolism, often referred to as physiological differentiation. Eventually cell multiplication ceases because of exhaustion of nutrient(s), accumulation of toxic waste products or density-dependent limitation of growth in monolayer cultures. There are means of prolonging the life of a batch culture, and thus increasing the yield, by various substrate feed methods.

(a)   Intermittently, by replacing a constant fraction of the culture with an equal volume of fresh medium.

(b)   Continuously, by slowing increasing the volume at a constant rate but withdrawing culture aliquots at intervals.

(c)   Perfusion, by the continuous addition of medium to the culture and the withdrawal of an equal volume of used (cell-free) medium. Perfusion can be open, that is the complete removal of medium from the system, or closed which is re-circulation of the medium, usually via a secondary vessel, back into the culture vessel. The secondary vessel is used to 'regenerate' the medium by gassing and pH correction.

All batch culture systems retain, to some degree, the accumulating waste products and have a fluctuating environment. All are suitable for both monolayer and suspension cells.

2.8.3 *Continuous-flow Culture*

This system gives true homeostatic conditions with no fluctuations of nutrients, metabolites or cell numbers. It depends upon medium entering the culture with a corresponding withdrawal of medium plus cells. It is thus only suitable for suspension culture cells, or monolayer cells growing on microcarriers. Continuous culture is described more fully in Section 4.6.

2.8.3 *Choice between Batch and Continuous-flow Culture*

Continuous-flow culture is the only system in which all cells are homogeneous, and can be kept homogeneous for long periods of time (months). This can be vital for physiological studies but may not be the most economical method for product generation. Production economics are calculated in terms of staff time, medium, equipment and downstream processing costs. Also taken into account are the complexity and sophistication of the equipment and process as this effects the calibre of the staff required and the reliability of the production process. Batch culture is expensive on staff time and culture ingredients because for every single harvest a sequence of inoculum build-up steps and then growth in the final vessel has to be carried out. Feeding routines for batch cultures can give repeated, but smaller, harvests and the longer a culture can be maintained in a productive state then the more economical the whole process becomes. However, to maintain a cell population at high levels for an extended period often needs large volumes of medium, and if one is using expensive growth factors then this may become the more expensive option. Maintenance of high cell yields, and therefore high product concentration, may be necessary to reduce downstream processing costs and these could outweigh medium expenses. Continuous-flow culture in the chemostat implies that cell yields are never maximal because a limiting growth factor is used to control the growth rate. If maximum yields are desired in this type of culture then the turbidostat option has to be used (see Section 4.6). Some applications, such as the production of a cytopathic virus, leave no choice other than batch culture.

3. MONOLAYER CULTURE

3.1 **Introduction**

Tissue culture flasks and tubes giving surface areas of $5-200$ cm^2 are familiar to all tissue culturists. The largest stationary flask routinely used in laboratories is the Roux bottle or disposable plastic equivalent which gives a surface area for cell attachment of $175-200$ cm^2 (depending upon type), needs $100-150$ ml medium and utilises $750-1000$ cc of storage space. This vessel will yield 2×10^7 diploid cells and up to 10^8 heteroploid cells. If one has to produce, for example, 10^{10} cells then over 100 replicate cultures are needed (i.e., manipulations have to be repeated 100 times). In addition the cubic capacity of incubator space needed is over 100 l. Clearly there comes a time in the scale-up of cell production when one has to use a more efficient culture system. Scale-up of anchorage dependent cells requires the reduction of the multiplicity of cultures, for more efficient use of manpower, and to increase significantly the surface area to volume ratio. In order to do this a very wide and versatile range of tissue culture vessels and systems have been developed. Many of these are shown diagrammatically in *Figure 6* and are described on the following pages. The methods with the most potential are those based on modifications to suspension culture systems because they allow a truly homogeneous, unit process, system with enormous scale-up potential, to be used. However, these systems should only be attempted if time and resources allow a lengthy development period.

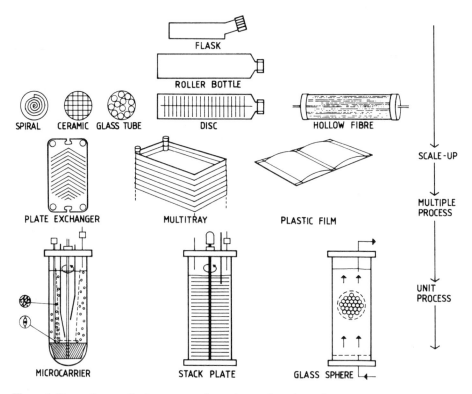

Figure 6. The scaling-up of culture systems for anchorage-dependent cells.

Although suspension culture is the preferred method for increasing capacity, monolayer culture has the following advantages.

(a)   It is very easy to change the medium completely and to wash the cell sheet before adding fresh medium. This is important in many applications when growth is carried out in one set of conditions, and product generation in another. A common requirement of a medium change is the transfer of cells from serum to serum-free conditions. The efficiency of media changing in monolayer cultures is such that a total removal of the unwanted compound can be achieved.

(b)   If artificially high cell densities are needed then these can be supported by using perfusion techniques. It is much easier to perfuse monolayer cultures because a fine filter system (to withhold cells) is not required.

(c)   Many cells will express a required product more efficiently when attached to a substrate.

(d)   The same apparatus can be used with different media:cell ratios which, of course, can be easily changed during the course of an experiment.

(e)   Monolayer cultures are more flexible because they can be used for all cell types. If a variety of cell types are to be used then a monolayer system might be a better investment.

It should be noted that the microcarrier system confers the advantages of (d) and (e) and to a lesser extent (b) to a suspension culture system. There are four main disadvantages of monolayer compared to suspension systems.

(a)   Difficult and expensive to scale up.

(b)   Utilise far more space.

(c)   Cell growth cannot be monitored so effectively due to the difficulty of sampling and counting an aliquot of cells.

(d)   It is more difficult to measure and control parameters such as pH and O_2 and to have homogeneity throughout.

In all cases microcarrier culture confers these advantages on a monolayer system.

3.2 Cell attachment

Animal cell surfaces and the traditional glass and plastic culture surfaces are negatively charged. Thus for cell attachment to occur, cross-linking with glycoproteins and/or divalent cations (Ca^{2+}, Mg^{2+}) is required. The glycoprotein most studied in this respect is fibronectin, a high molecular weight (220 000) compound synthesised by many cells and present in serum and other physiological fluids. Although cells can presumably attach by electrostatic forces alone it has been found that the mechanism of attachment is similar, whatever the substrate charge (7). The important factor is the net negative charge and surfaces, such as glass and metal, which have high surface energies are very suitable for cell attachment. Organic surfaces need to be wettable and negative and this can be achieved by chemical treatment (e.g., oxidising agents, strong acids) or physical treatment (e.g., high voltage discharge, u.v. light, high energy electron bombardment). One or more of these methods are used by suppliers of tissue culture grade plastics. The result is to increase the net negative charge of the surface (by forming negative carboxyl groups for example) for electrostatic attachment. Surfaces may also be coated to make them suitable for cell attachment. Collagen and poly-amino acids are the most commonly used agents. A tissue culture grade of collagen can be purchased (Vitrogen 100, Flow Laboratories) which saves a tedious preparation procedure using rat tails. The usefulness of collagen as a growth surface is also demonstrated by the availability of collagen-coated microcarrier beads (Cytodex-3, Pharmacia Fine Chemicals).

3.2.1 *Surfaces for Cell Attachment*

(a) *Glass.* Alum-borosilicate glass (e.g., Pyrex) is the preferred type because soda-lime glass releases alkali into the medium and should be de-toxified (by boiling in weak acid) before use. Glassware after repeated use can become less efficient for cell attachment but this efficiency can be regained by treatment with 1 mM magnesium acetate. After several hours soaking at room temperature the acetate is poured away, the glassware rinsed with distilled water and autoclaved.

(b) *Plastics.* Polystyrene is the most used plastic for cell culture but polyethylene, polycarbonate, perspex, PVC, Teflon, cellophane and cellulose acetate are all suitable when pre-treated correctly.

(c) *Metals.* Both stainless steel and titanium are suitable for cell growth because they are relatively chemically inert, but have a suitable high negative energy. There are many grades of stainless steel and care has to be taken in choosing those which do not leak toxic metallic ions. The most common grade to use for culture applications is 316, but 321 and 304 may also be suitable. Stainless steel should be acid washed (10% nitric acid, 3.5% hydrofluoric acid, 86.5% water) to remove surface impurities and inclusions caused during cutting.

3.3 **Scaling-Up: Step 1, Roller Bottle**

The aims of scaling-up are to maximise the available surface area for growth and to minimise the volume of medium and headspace, while optimising cell numbers and productivity. Stationary cultures have only one surface available for attachment and growth and consequently, they need a large medium volume. The medium volume can be reduced by rocking the culture or, more usually, by rolling a cylindrical vessel. The roller bottle has nearly all its internal surface available for cell growth although only 15 − 20% is covered by medium at any one time. Rotation of the bottle alternately subjects the cells to medium and air as compared with the near anaerobic conditions in a stationary culture. This method reduces the volume of medium required but still requires a considerable headspace volume to maintain adequate oxygen and pH levels. The scale-up of a roller bottle requires that the diameter is kept as small as possible. The surface area can be doubled by doubling the diameter or the length. The first option increases the volume (medium and headspace) 4-fold, but the second option only 2-fold.

The only means of increasing the productivity of a roller bottle and decreasing its volume is by using a perfusion system, originally developed by Kruse (8), and marketed by New Brunswick Scientific Company and Bellco Glass Inc. (autoharvester). This is an expensive option as an intricate revolving connection has to be made for the supply lines to pass into the bottle. However, cell yields are considerably increased and extensive multilayering takes place.

3.3.1 *Experimental Protocol for using Standard Disposable Roller Cultures*

The following procedure is based on the use of a 1400 cm² (23 × 12 cm) plastic disposable bottle (Corning or Becton Dickinson).

(i)   Add 300 ml of growth medium.

(ii) Add 1.5 × 10⁷ cells.

(iii) Roll at 12 r.p.h. at 37°C for 2 h, to allow an even distribution of cells during the attachment phase.

(iv) Decrease the revolution rate to 5 r.p.h., and continue incubation.

(v)   Examine cells under an inverted microscope using a long working distance objective.

(vi) Harvest cells when visibly confluent (5 – 6 days) by removing the medium, adding trypsin (0.25%) and rolling. Yields will be very similar to those obtained in flasks assuming enough medium was added. This method has the advantage of allowing the medium volume:surface area ratio to be altered easily. Thus after a growth phase the medium volume can be reduced to get a higher product concentration.

3.4 Scaling-Up: Step 2, Roller Bottle Modifications

The roller bottle system is still a multiple process, and thus inefficient in terms of manpower and materials. To increase the surface area within the volume of a roller bottle the following vessels have been developed (see also *Figure 6*).

3.4.1 *Spiral Film*

The Sterlin bulk cell culture vessel is 25 × 10 cm and is fitted with a cartridge consisting of a spirally wound plastic film giving a surface area of 8500 cm². It is not a true roller bottle in that it is only rolled during the cell attachment phase (3 – 16 h) at 2 – 3 r.p.h. Once the cells are attached the vessel is stood upright and, because the vessel is filled with medium (1.8 l) it has to be aerated by sparging. The system has proved successful for heteroploid cell lines, such as BHK and 3T3, with yields of up to 1 – 2 × 10⁹ from a 1 × 10⁸ inoculum. The unit has not proved so satisfactory for diploid cells. It is crucial to get an even distribution of the cell inoculum throughout the spiral, otherwise very uneven growth and low yields are achieved. Cell growth can only be visualised on the outside of the spiral and this can be misleading if the cell distribution is uneven.

3.4.2 *Glass Tubes*

A small scale example is the Bellco-Corbeil Culture System (Bellco Glass Inc.). This was developed by Corbeil (9). A roller bottle is packed with a parallel cluster of small glass tubes (separated by silicone spacer rings). Three versions are available giving surface areas of 5 × 10³, 1 × 10⁴ and 1.5 × 10⁴ cm². Medium is perfused through the vessel from a reservoir. The method is ingenious in that it alternately rotates the bottle 360° clockwise, and then, 360° anticlockwise. This avoids the use of special caps for the supply of perfused medium.

An example of its use is the production of 3.2 × 10⁹ Vero cells (2.3 × 10⁵/cm²) over 6 days using 6.5 l medium (perfused at 50 ml/min) in the 10 000 cm² version.

A similar design is the Gyrogen (Chemap A.G.). Six sizes are available ranging from 0.52 to 34 cm² in surface area. The smaller units are in a glass vessel but the larger ones are mounted in a stainless steel cylinder with provisions for tilting, environmental control, microscopical examination and *in situ* sterilisation.

3.4.3 *Plates*

These systems are now largely unavailable but are described because the principle still exists in larger equipment and they can be easily constructed in the laboratory. A rola cartridge was supplied by Linbro which contained 40 discs arranged in parallel on a centre spindle. This gave a surface area of 6300 cm² in a 12 × 22 cm roller tube. The unit was half filled with medium and used as a conventional roller bottle.

A much larger version using titanium discs, which gave a total surface area of 1.7 m², used to be available from Biotec Inc. A disadvantage of this unit was that confluent cell sheets were prone to sloughing off and is obviously why the systems are no longer used, although the principle has been transferred to the stack-plate type fermentors (Section 3.6.2). The main disadvantage of vertical plate systems is getting even cell attachment with only half-filled bottles.

3.5 Scaling-Up: Step 3, Large Capacity Stationary Cultures

3.5.1 *Multitray Unit (A/S NUNC)*

The standard multitray unit comprises ten chambers, each having a surface area of 600 cm², fixed together vertically and supplied with interconnecting channels. This enables all operations to be carried out once only for all chambers. It can thus be thought of as a flask with a 6000 cm² surface area using 2 l medium and taking up a total volume of 12 500 cc. In practice this unit is covenient to use and produces good results, analogous to plastic flasks. It is made of tissue culture grade polystyrene and is disposable. One of the disadvantages of the system can be turned to good use. In practice it is difficult to wash out all the cells after harvesting with trypsin etc. However, enough cells remain to inoculate a new culture when fresh medium is added. Given good aseptic technique this disposable unit can be re-used 3 − 4 times. The system is used commercially for interferon production (by linking together multiples of these units) (10). In addition units are also available giving 1200 and 24 000 cm².

3.5.2 *Hollow fibre Culture*

Bundles of synthetic hollow fibres offer a matrix analogous to the vascular system and allow cells to grow in tissue-like densities. Hollow fibres are usually used in ultrafiltration, selectively allowing passage of macromolecules through the spongy fibre wall while allowing a continuous flow of liquid through the lumen. When these fibres are enclosed in a cartridge and encapsulated at both ends, medium can be pumped in and will then perfuse through the fibre walls, which provide a large surface area for cell attachment and growth.

Culture chambers based on this principle are available from Amicon Corporation (Vitafibre). The capillary fibres are made of acrylic polymer, are 350 μm in diameter with 75 μm walls. The pores through the internal lumen lining are available with molecular weight cut offs between 10 000 and 100 000. It is difficult to calculate the total surface area available for growth but units are available in various sizes and these give a very high surface area to culture volume ratio (in the region of 30 cm^2/cc). Up to 10^8 cells/ml have been maintained in this system.

3.5.3 *Opticell Culture System*

A new culture system has recently been introduced by K.C. Biological which uses a ceramic material for cell growth. The system consists of a cylindrical ceramic cartridge (available in surface area between 4.25 and 18 m^2) with 1 mm^2 square channels running lengthwise through the unit. A medium perfusion loop to a reservoir, in which environmental control is carried out, completes the system. It provides a large surface area:volume ratio (40:1) and preliminary investigations have shown its suitability for virus and cell surface antigen production.

3.4.4 *Plastic Bags (11)*

Bags made of fluoro-ethylene-propylene copolymer (FEP-Teflon, Du Pont) are biologically inert but are very gas-permeable. Thus a bag of 5 × 30 cm filled with medium 2 − 10 mm deep and cells, can be placed in an incubator. Oxygen supply is sufficient (the cells grow on the inside of the bag and are thus separated by only 25 μm from air) to allow high cell densities to be obtained. The culture is rocked or rotated to keep the medium homogeneous. Cells can be harvested by conventional trypsination or mechanically by folding and stretching the bags.

3.5.5 *IL-410 Plastic Film Propagator (12)*

This method uses the same material as above (FEP-Teflon), but instead of bags it is presented as a long tube and wrapped with a spacer around a reel. Medium is pumped through the tubing and can be either recirculated after passing through a reservoir, or discarded. Gas exchange occurs between the medium and the incubator air space through the wall of the tubing. Lengths of tubing up to 10 m, giving a surface area for growth of 25 000 cm^2 have been used.

3.5.6 *Plate Heat Exchanger*

Plate heat exchangers offer a large surface area of stainless steel with adequate provision for circulating medium on one side and water at 37°C on the other side of the growth surface. This system is available from AVP in units of 2 m^2 (13).

3.6 Scale-Up: Step 4, Fermenter Systems

There are basically three systems which fit into fermentation (suspension culture) apparatus (see *Figure 6*):

(a)   Cells stationary, medium moves (e.g., glass bead reactor).
(b)   Heterogeneous mixing (e.g., stack plate reactor).
(c)   Homogeneous mixing (e.g., microcarrier).

Table 2. Properties of 3 mm Glass Beads.

Weight (g)	0.0375
Surface area (cm²)	0.283
Volume (cc)	0.015
Packed volume (cc)	0.028
Thus a culture of 10 000 cm² =	
35 336	Glass beads
990 ml	Total volume
460 ml	Medium in culture
21 cm²/ml	Medium
20 × 8 cm	Typical culture size

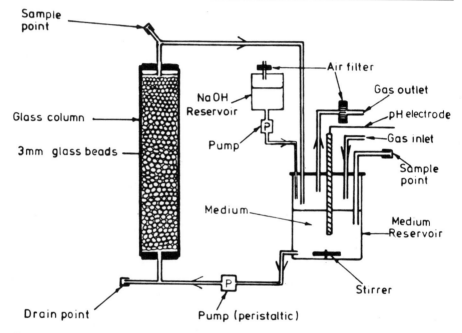

Figure 7. A glass bead culture system.

3.6.1 Bead Bed Reactor

The use of a packed bed of 3 – 5 mm glass beads, through which medium is continually perfused, for growing cells has been reported by a number of investigators since 1962. The potential of the system for scale-up was demonstrated by Whiteside and Spier (14) who use a 100 l capacity system for the growth of BHK21 cells (and FMDV).

Spheres of 3 mm pack sufficiently tightly to prevent the packed bed from shifting, but allow sufficient flow of medium through the column so that fast flow rates, which would cause mechanical shear damage, are not needed. The physical properties of glass spheres are given in *Table 2* and these data allow the necessary culture parameters to be calculated. A system which can be constructed in the laboratory is illustrated in *Figure 7*. Medium is transferred by a peristaltic pump in this example, but an air lift driven system is also suitable and gives better oxygenation. Medium can be passed either up or down the column with no apparent difference in results.

Experimental procedure

(i)   Prepare sufficient media at the rate of $1-2$ ml/10 cm^2 culture surface area.

(ii)   Circulate media between reservoir and vessel until the system is equilibrated with regard to temperature and pH (allow sufficient time for the glass beads to warm up to 37°C).

(iii)   Add cell inoculum ($\sim 1 \times 10^4$/cm^2) to a volume of medium equal to the void volume.

(iv)   Drain the bead column into the reservoir, mix the cells thoroughly and add to the bead column.

(v)   Clamp off all tubing and allow the cells to attach ($3-8$ h depending on cell type).

(vi)   Once cells are attached start perfusing the medium, initially at a slow rate (1 linear cm/10 min). Visibly inspect the effluent medium for cloudiness indicating that cells have not attached (stop perfusing if this is the case).

(vii)   After 24 h this rate can be increased but should be kept below 5 cm/min otherwise cells may be stripped off (especially mitotic cells). The pH readout at the exit end and a comparison of glucose concentrations in the input and exit sampling points will indicate whether the flow rate is sufficient.

(viii) Monitor cell growth by glucose determinations. Glucose utilisation rates should be found empirically for each cell and medium but a rough indication is $2-5 \times 10^8$ cells produced/g glucose.

(ix)   When the culture is assumed confluent drain off the medium, wash the column with buffer, add the void volume of trypsin/versene (or pronase etc.) and allow to stand for $15-30$ min. Cell harvesting can be accelerated by occasionally releasing the trypsin from the column into the harvest vessel and pushing it back into the culture again. Alternatively feed gas bubbles through the column (only very slowly otherwise the beads will shift and damage the cells).

(x)   The culture vessel should be washed immediately after use otherwise it becomes very difficult to clean. Use a detergent, such as Decon, and circulate continuously through the culture system, followed by tap water and then several changes of distilled water. Autoclave moist if control probes (O$_2$, pH) are used. Use Pyrex glass beads in preference to the standard laboratory beads which are made of soda glass.

3.6.2 *Heterogeneous Reactor*

This culture is based on the same principle as the disc roller cultures described previously (Section 3.4.2). Circular glass or stainless steel plates are fitted vertically, $5-7$ mm apart, on a central shaft. This shaft may be stationary, with an airlift pump for mixing, or revolving around a vertical (6 r.p.h.) or horizontal ($50-100$ r.p.m.) axis. The multisurface propagator (15) was used at sizes ranging from 7.5 to 200 l, giving a surface area of up to 2×10^5 cm^2. The author's experience is solely with the horizontal stirred plate type of vessel (*Figure 6*, Stack plate), which is easy to use and has been successful for both heteroploid and human diploid cells. The main disadvantage with

53

this type of culture is the high medium volume to surface area (1 ml to $1-2$ cm²). This cannot be altered with the horizontal types although it can be halved with the vertically revolving discs.

3.6.3 *Homogeneous Systems (Microcarrier)*

When cells are grown on small spherical carriers they can be treated as a suspension cell, and advanced fermentation technology processes and apparatus can be utilised. The method was initiated by Van Wezel (16), who used dextran beads (Sephadex A-50). These were not entirely satisfactory as the charge of the beads was unsuitable and also possibly due to toxic effects. However, much developmental work has since resulted in many suitable microcarriers being commercially available (*Table 3*).

The authors experience has largely been with the Cytodex (Pharmacia) microcarriers and much of the following discussion is based on this particular product. The choice of this microcarrier was based on: (a) a preference for a dried product which could be accurately weighed and then prepared *in situ*; and (b) the fact that with a density of only 1.03 g/cc this product could be used at concentrations of up to 15 g/l (90 000 cm²/l). However this preference does not detract from the quality of other microcarriers, most of which have been used with equal success.

Table 3. Comparison of Microcarriers.

	Cytodex 1	Cytodex 2	Cytodex 3	Superbeads	Biocarrier
Manufacturer	Pharmacia	Pharmacia	Pharmacia	Flow Labs	BioRad
Material	DEAE - Sephadex	DEAE - Sephadex	collagen (60 mg/cm²)	DEAE - Sephadex	Polyacrylamide
Specific gravity	1.03	1.04	1.04	1.03	1.04
Shape	Spherical	Spherical	Spherical	Spherical	Spherical
Size (μdiam)	$160-230$	$115-200$	$130-210$	$150-200$	$120-180$
Surface area cm²/g	6000	5500	4600	6000	5000
Price/m² (relative units)	2.5	2.8	4.0	4.0	4.0
Formulation	Dry Non-sterile	Dry Non-sterile	Dry Non-sterile	Solution Sterile	Dry Non-sterile

	Biosilon	Cytosphere	Gelibead	Cellulose	Acrobead
Manufacturer	Nunc	Lux	KC Biological	Whatman	Galil
Material	Polystyrene	Polystyrene	Gelatin	DEAE cellulose	Various
Specific gravity	1.05	1.04	1.04	?	1.04
Shape	Spherical	Spherical	Spherical	Cylindrical	Spherical
Size (μdiam)	$160-300$	$160-230$	$115-235$	Very variable	$150-150$
Surface area cm²/g	255	250	3800	$2-4000$	5000
Price/m² (relative units)	11	45	3.5	1.5	—
Formulation	Dry Sterile	Dry Sterile	Dry Non-sterile	Moist Non-sterile	Dry Non-sterile

(i) *Culture apparatus.* A spinner vessel is not suitable unless the stirring system is modified. Large surface area paddles are needed. These are supplied in the Belco μ spinner vessels, but can be easily constructed out of a silicone rubber sheet and attached to the magnet with plastic ties. The advantage of constructing one within the laboratory is that the blades can be made to incline 20 − 30° from the vertical, thus giving much greater lift and mixing than vertical blades (*Figure 8*). Microcarrier cultures are stirred very slowly (maximum 75 r.p.m.) and it is essential to have a good quality magnetic stirrer that is capable of giving a smooth stirring action in the range of 20 − 100 r.p.m. Never use a stirrer mechanism that has moving surfaces in contact with each other in the medium, otherwise the microcarriers will be crushed. Thus, stirrers which revolve on the bottom of the vessel are unsuitable. As mixing, and thus mass transfer, is so poor in these cultures the depth of medium should not exceed the diameter by more than a factor of two unless oxygenation systems or regular medium changes are performed. In the basic culture systems, media changes have to be carried out at frequent intervals. It is worthwhile, therefore to make suitable connections to the culture vessel to enable this to be done conveniently *in situ*. This will pay dividends in reducing the chances of contaminating the culture. A simplified culture set up is shown in *Figure 9*.

(ii) *Initiation of the culture.* Many of the factors already discussed in Section 2 are critical when initiating a microcarrier culture. Microcarriers are spherical and cells will always attach to an area of minimum curvature. Therefore a microcarrier surface can never be ideal however suitable its chemical and physical properties. Ensure the media and beads are at a stable pH and temperature, and inoculate the cells (from a

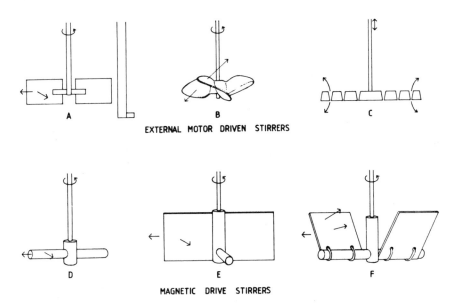

Figure 8. Types of impellers for growing suspension and microcarrier cells. (**A**) Flat disc turbine, d2:d1 = 0.33, radial flow, turbulent mixing; (**B**) Marine propeller, d2:d1 = 0.33, axial flow, turbulent mixing, angle 25°; (**C**) Vibro-mixer; (**D**) Stirrer bar, d2:d1 = 0.6, radial flow, laminar mixing; (**E**) Vertical and (**F**) Angled (25°) Paddle, d2:d1 = 0.6 − 0.9, axial and radial flow, laminar and turbulent mixing. (d2 = diameter of impeller:d1 = diameter of vessel).

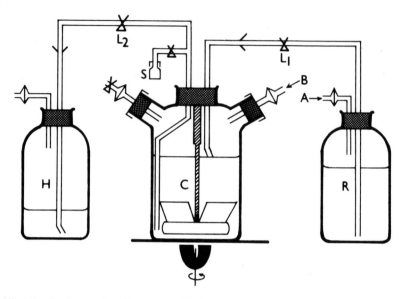

Figure 9. A simple microcarrier culture system allowing easy medium changing. To fill the culture (c) open clamp (L1) and push the medium from the reservoir (R) using air pressure (at A). To harvest, stop the stirring for 5 min, open L2 and push the medium from C to H using air pressure at B. S is a sampling point.

logarithmic, not stationary culture) into a third of the final medium volume. This increases the chances of cells coming into contact with the microcarriers. Microcarrier concentrations of $2-3$ g/l should be used. Higher concentrations need environmental control or very frequent medium changes. Historically it was considered best to allow attachment in a stationary culture and to stir for 30 sec every 30 min. However, this always gave rise to uneven numbers of cells per bead and now, with the vast improvements in microcarrier quality, stirring can start immediately at the lowest possible setting that gives complete mixing ($10-25$ r.p.m.).

After the attachment period ($3-8$ h) slowly top up the culture to the working volume and increase the stirring rate to maintain completely homogeneous mixing. If these conditions are adhered to, and there are no changes in temperature and pH then all cells which grow on plastic surfaces should readily initiate a microcarrier culture.

(iii) *Maintenance of the culture*. It is very easy to monitor the progress of cell growth in a microcarrier culture. Samples can be easily removed, cell counts (by nuclei counting) and glucose determinations carried out, and the cell morphology examined. As the cells grow so the beads become heavier and need an increased stirring rate. After 3 days or so the culture will become acidic and need a medium change. Again this is an extremely easy routine, turn off the stirrer, allow the beads to settle for 5 min, and decant off as much medium as desired. Top up gently with fresh medium (prewarmed to 37°C) and restart the stirring.

(iv) *Harvesting*. It is very difficult to harvest many cell types from microcarriers unless the cell density on the bead is very high. Incidentally, do not expect cells to multilayer on microcarriers, this hardly ever happens. Harvesting can be attempted by draining off the medium, washing the beads at least once in buffer and adding the desired en-

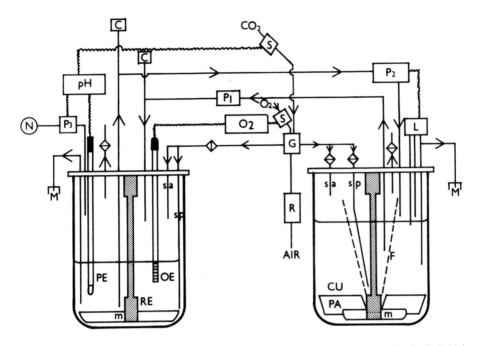

Figure 10. A closed loop perfusion system with full environmental control for the growth of cells in high density microcarrier cultures. CU, culture vessel; RE, reservoir; C, connector for medium changes, harvesting, etc.; F, filter; G, gas blender; L, level controller; M, sampling device; m, magnet; N, alkali (NaOH) reservoir; OE, oxygen electrode; PE, pH electrode; P, pumps; 1, medium to reservoir (continuous), 2, medium to culture (controlled by L), 3, alkali to reservoir (controlled by pH meter); R, flowmeter; SA, gas supply for surface aeration; SP, gas supply for sparging; S, solenoid valve.

zyme. Stir the culture fairly rapidly (75 – 125 r.p.m.) for 20 – 30 min. If the cells detach, a high proportion can be collected by allowing the beads to settle out for 2 min and then decanting of the supernatant. For a total harvest pour the mixture into a sterilised sintered glass funnel (porosity grade 1). The cells will pass through the filter, but not the microcarriers.

Recent additions to the microcarrier range make harvesting far simpler. These are the collagen coated Cytodex-3 beads from which cells can be released with collagenase, and the Corning gelatin beads which are solubilised with trypsin and/or ethylenediamine-tetraacetic acid (EDTA).

(v) *Scaling-up of microcarrier culture.* Scaling-up can be achieved: (a) by increasing the microcarrier concentration; and (b) by increasing the culture size. If option (a) is used nutrients and oxygen are very rapidly depleted and the pH falls to non-physiological levels. Medium changes are not only tedious but provide rapidly changing environmental conditions. Perfusion, either to waste or a closed loop, must be used to achieve cultures with high microcarrier concentration. This can only be brought about by an efficient filtration system so that medium without cells and microcarriers can be withdrawn at a rapid rate. The only satisfactory means of doing this is with the type of filtration system illustrated in *Figure 10*. This is constructed of a stainless steel mesh with an absolute pore size in the 60 – 120 micron range. Attachment to the stirrer shaft means that a large surface area filter can be used and the revolving action discourages cell attachment and clogging (17).

However scaling-up is achieved, oxygen limitation is the chief factor to overcome. This is an especially difficult problem in microcarrier culture because stirring speeds are low (it was pointed out in Section 2.7 that the stirring speed has to be above 100 before it significantly affects the aeration rate). Sparging cannot be used as microcarriers get left literally high and dry above the medium level due to the foaming it causes. The perfusion filter previously described, however, does allow sparging into that part of the culture in which no beads are present. Unfortunately, this means that the most oxygenated medium in the culture is removed by perfusion but at a sparging rate of 10 cc of oxygen/l considerable diffusion of oxygenated medium occurs into the main culture. Thus oxygen delivery to a microcarrier culture can utilise the following systems:

(a)   Surface aeration.
(b)   Increasing the perfusion rate of fully oxygenated medium from the reservoir.
(c)   Sparging into the filter compartment (18).

These three systems are used by the author in the system illustrated in *Figure 10* to run cultures at up to 15 g/l Cytodex-3 in culture vessels between 2 and 20 l working volume. A typical experiment is shown in *Figure 11*.

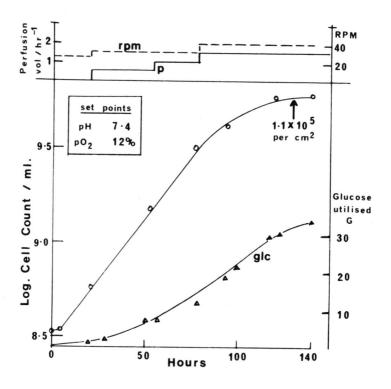

Figure 11. Growth of GPK epithelial cells on Cytodex-3 (10 g/l) in the 10 l culture system described in *Figure 10*.

Table 4. Comparison of the Medium Volumes and Total Volumes of a Range of Monolayer Culture Systems.

	Surface area cm²	Surface area (cm²)	
		per cc of medium	per cc of culture volume
Roux	175 – 200	2	0.2
Roller	750 – 1500	4	0.3
— Spiral film	8500	5	5
— Glass tubes	5000 – 340 000	2	4[b]
— Plates	6300 – 17 000	6 – 10	3 – 5
Multitray	6000 – 24 000	2.5	0.5
Hollow fibres	1000 – 5000	(30)	(25)[b]
Opticell	42 500 – 180 000	40	35[b]
Plastic bags	300 → [a]	5	5
Plastic roll	25 000	25	20
Glass bead propagator	2×10^6 → [a]	20	19[b]
Multisurface propagator	2.5×10^5	1.1	1.0
Stack plate fermenter	3.5×10^5 → [a]	1.7	1.3
Plate exchanger	2×10^4 → 1.5×10^7	6.7	3[b]
Microcarrier 5 g/l	30×10^6 → [a]	30	24[b]
15 g/l	2×10^6 → [a]	90	75[b]

[a]At known maximum culture systems, not the maximum potential.
[b]Plus reservoir and control equipment.

(vi) *Summary of microcarrier culture.* Microcarrier cultures are used commercially for vaccine and interferon production in fermenters up to 1000 l. These processes use heteroploid or primary cells. Unfortunately, results with human diploid cells are far too inconsistent to recommend this procedure for cultures of this type. Reasonable performance with these cells is only possible in laboratory scale equipment (<20 l) where far more individual attention can be given. One of the problems of using very large scale cultures is that the required seed inoculum gets progressively larger. Harvesting of cells from large unit scale cultures is not always very sucessful, although the availability of collagen and gelatin microcarriers eases this problem considerably. Microcarriers cost in the region of £1200 – 1500/kg. Some consideration must be given to washing and re-using the beads but this, of course, is not possible with the gelatin- or collagenase-treated beads.

3.7 Summary

A wide range of commercially available and laboratory made equipment has been reviewed. This should enable a choice to be made depending upon the amount and type of cells or product needed and the financial and manpower sources available. To review the choices available *Figure 6* and *Table 4* should be consulted.

4. SUSPENSION CULTURE

As indicated previously in this chapter, suspension culture is the preferred method for scaling-up cell cultures. Some cells, especially those of haemopoietic derivation, grow best in suspension culture, others can be adapted or selected, while a few, for example human diploid cell strains (W1-38, MRC-5), will not survive in suspension at all. A further factor that dictates whether suspension systems can be used is that some cellular products are only expressed when the cell exists in a monolayer or if cell to cell contact is established (e.g., for the spread of intracellular viruses through a cell population).

4.1 Adaptation to Suspension Culture

Different cell lines vary in the ease with which they can be persuaded to grow in suspension. For those that have the potential, there are two basic procedures that can be used to generate a suspension cell line from an anchorage-dependent one.

4.1.1 Selection

This method, as demonstrated by the derivation of the LS cell line from L-929 cells by Paul and Struthers, and the HeLa-S3 clone from HeLa cells, depends on the persistence of loosely attached variants within the population. A confluent monolayer culture is lightly tapped, or the medium gently swirled, the medium decanted from the culture and cells in suspension recovered by centrifugation. Cells have to be collected from many cultures to provide a sufficient number to start a new culture at the required inoculum density (at least 2×10^5/ml). This procedure has to be repeated many times over a long period because many of the cells that are collected are in the mitotic phase, rather than potential suspension cells. This is because cells round-up and become very loosely attached to the substrate during mitosis. Eventually it is possible with some cell lines to derive a viable cell population that divides and grows in static suspension, just resting on the substrate rather than attaching and spreading out.

4.1.2 Adaptation

This method probably works the same way as the selection procedure except that a selection pressure is exerted on the culture while it is maintained in suspension mechanically. However, with many cell lines the relatively large number that become anchorage-independent suggests it is more than just selection of a few variant cells. The following method, used successfully by the author for many cell lines, is based on the one published by Capstick *et al.* (19) for BHK21 cells.

(i)   The cell suspension is prepared by detaching monolayer cells with trypsin (0.1%) and versene (0.01%).

(ii)   At least two, but preferably three or more, spinner vessels are made ready by adding the growth medium (calcium- and magnesium-free modification of any recognised culture medium).

(iii)   Add cells at a concentration of at least 5×10^5/ml and commence stirring [the lowest rate which keeps the cells homogeneously mixed (e.g., 250 r.p.m.)].

(iv)   Sample and carry out viable cell counts every 24 h.

(v)   Every 3 days remove the medium from the culture, centrifuge (800 r.p.m.) and resuspend the cells in fresh medium at a density of at least 2.5×10^5/ml. Depending upon the degree of cell death, cultures will almost certainly have to be amalgamated to keep the cell density at the required level.

(vi)   If, at a medium change, there is significant attachment of cells to the culture vessel, especially at the air-medium interface, then add a trypsin (0.01%)/versene

(0.01%) mixture to the empty vessel. Stir at 37°C for 30 min and recover the detached cells by centrifugation. If the cells in suspension are badly clumped, they too can be added to the trypsin/versene solution. This treatment is usually necessary at the first and second media changes, but very rarely after this point. If cell clumping persists, then add 50 μg/ml trypsin or Dispase (Boehringer) to the growth medium.

(vii)   Successful adaptation is recognised initially by an increase in cell numbers after a media change, and subsequently by getting consistent cell yields in a given period of time (indicating a constant growth rate).

(viii) It is usually necessary to maintain the newly established cell strain in stirred suspension culture as reversion to anchorage-dependence can occur in static cultures. Sometimes cells will adhere to the substrate without becoming completely spread out and continue to grow and divide with a near-spherical morphology.

4.2 Static Suspension Culture

Many cell lines will grow as a suspension in a culture system used for monolayer cells (i.e., with no agitation or stirring). Cell lines that are capable of this form of growth include the many lymphoblast lines (e.g., MOLT, RAJI), hybridomas and some non-haemopoietic lines, such as the LS cells described in the previous section. However, with the latter types of cell there is always the danger of reversion to a monolayer (e.g., a small proportion of the LS cell line always attaches and is discarded at each sub-cultivation). Static suspension culture is unsuitable for scale-up for reasons already stated for monolayer culture.

4.3 Small-scale Suspension Culture

For the purposes of this chapter, small scale culture is defined as under 2 l. This is an entirely arbitrary definition, but is made on the basis that above 2 l additional factors apply. The conventional laboratory suspension culture is the spinner flask, so called because it contains a magnetic bar as the stirrer and this is driven from below the vessel by a magnetic stirrer. Details of some of the readily available stirrer vessels, together with suppliers and available size range, are given in *Figure 12*. It is important to get a good quality magnetic stirrer so that the magnetic field is strong enough to turn the magnet smoothly over the prescribed range of stirring speeds (50 – 500 r.p.m.) and is reliable over long periods of use. Additional requirements are that it will not get too hot (as the culture vessel sits directly on top) and that a tachometer is included that gives a true reading of stirring speed. Always check that the stirrer will re-start from all required settings after an interruption in power.

4.4 Scaling-up Factors

In scale-up, both the physical and chemical requirements of cells have to be satisfied. The chemical factors require environmental monitoring and control to keep the cells in the proper physiological environment. These factors, which include oxygen, pH, medium components and removal of waste products, are described in Section 2.

Physical parameters include the configuration of the bioreactor and the power supplied to it. The function of the stirrer impeller is to convert energy (measured as kW/m³)

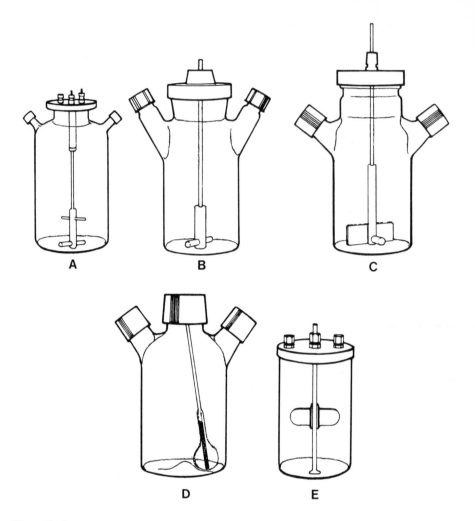

Figure 12. Commercially available spinner cultures. **(A)** LH Fermentation Biocul (1 – 20 l); **(B)** Bellco and Wheaton spinner flasks (25 ml – 2 l); **(C)** Bellco μ spinner (25 ml – 2 l); **(D)** Techne (0.25 – 5 l); **(E)** Techne Cytostat (1 l).

into hydrodynamic motion in three dimensions (axial, radial and tangential). The impeller has to circulate the whole liquid volume and to generate turbulence (i.e., it has to pump and to mix) in order to: create a homogeneous blend; to keep cells in suspension; to optimise mass transfer rates between the different phases of the system (biological, liquid and gaseous); and to facilitate heat transfer.

Good mixing becomes increasingly difficult with scaling-up and the power needed to attain homogeneity can cause problems. The energy generated at the tip of the stirrer blade is a limiting factor as it gives rise to a damaging shear force. Shear forces are created by fluctuating liquid velocities in turbulent areas. The factors which affect this are: impeller shape (this dictates the primary induced flow direction) (*Figure 8*); the impeller to vessel diameter; and the impeller tip speed (a function of rotation rate and diameter). The greater the turbulence the more efficient the mixing, but a compromise has to be reached so that cells are not damaged. Large impellers running at low speeds give a low shear force and high pumping capacity, whereas smaller impellers need high stirring speeds and have high shear effects. This is the basis for helical type stirrers (20).

Magnetic bar stirring gives only radial mixing and no lift or turbulence. The marine impeller is more effective for cells than the flat blade turbine type impellers found in many bacterial systems, as it gives better mixing at low stirring speeds. If the cells are too fragile for stirring, or if sufficient mixing cannot be obtained without causing unacceptable shear rates then an alternative mixing system may have to be used. Pneumatic energy, for example mixing by air bubbles (e.g., airlift fermenter) or hydraulic energy (e.g., medium perfusion), can be scaled-up without proportionally increasing the power. To improve the efficiency of mechanical stirring, the design of the stirrer paddle can be altered (e.g., as described for microcarrier culture), or multiple impellers can be used. Some examples are given in *Figure 8*.

A totally different stirring concept is the Vibro-mixer. This is a non-rotating agitator which produces a stirring effect by a vertical reciprocating motion with path of $0.1-3$ mm at a frequency of 50 Hz. The mixing disc is fixed horizontally to the agitator shaft and conical shaped holes in this disc cause a pumping action to occur as the shaft vibrates up and down. The shaft is driven by a motor which operates through an elastic diaphragm; this also provides a seal at the top of the culture. A fermentation system using this principle is available commercially (Vibro-Fermenter, Chemap). The advantages of this system are the greatly reduced shear forces, random mixing, reduced foaming and reduced energy requirement, especially for scaling-up.

The significant effect on vessel design of moving from a magnetic stirrer to a direct drive system is the fact that the drive has to pass through the culture vesel. This means some complexity of design to ensure a perfect aseptic seal while transferring the drive, complete with lubricated bearings, through the bottom or top plate of the vessel.

Scaling-up cannot be a proportional event; one cannot convert a 1 l reactor into a 1000 l reactor simply by increasing all dimensions by the same amount. The reasons for this are mathematical, doubling the diameter increases the volume 3-fold and this affects the different physical parameters in different ways. One factor to be taken into consideration is the height to diameter ratio as this is one of the most important fermenter design parameters. In sparged systems the taller the fermenter in relation to its diameter the better, as the air pressure will be higher at the bottom (increasing the oxygen solubility) and the residence time of bubbles longer. However, in non-sparged systems which are often used for animal cells and rely on surface aeration, the surface area to height ratio is more important and a 1:1 ratio should be maintained.

Mass transfer between the culture phases has been discussed in relation to oxygen (Section 2.7). It is this characteristic of scaling-up which demands the extra sophistication in culture design to maintain a physiologically correct environment. This sophistication includes impeller design, oxygen delivery systems, vessel geometry, perfusion loops, or a completely different concept in culture design to the stirred reactor.

4.5 **Stirred Fermenters**

The move from externally driven magnetic spinner vessels to fermenters capable of scale-up from 1 l to 1000 l and beyond, has the following consequences at some stage.

(a)   Change from glass to stainless steel vessels.

(b)   Change from a mobile to a static system: connection to steam for *in situ* sterilisation; requirement for water-jacket or internal temperature control; and need for a seed vessel, a medium holding vessel and downstream processing capability.

(c)   Greater sophistication in environmental control systems to meet the increasing mass transfer requirement.

In practice, the maximum size for a spinner vessel is 20 l. Above this size there are difficulties in handling and autoclaving, as well as the difficulty of being able to agitate the culture adequately. Fermenters with motor driven stirrers are available from 500 ml, but these are chiefly for bacterial growth. It is a significant step to move above the 10 — 20 l scale as the cost of the equipment is significant (e.g., £15 000 — 20 000 for a complete 35 l system) and suitable laboratory facilities are required (steam, drainage, etc.). There is a wide range of vessels available from the various fermenter suppliers but most of them were developed for bacterial growth and are not ideal for most animal cells. However, fermenters more suitable for animal cells are now becoming available and include some of the following modifications.

(i)   suitable impeller (e.g., marine);

(ii)   no baffles;

(iii)   curved bottom for better mixing at low speeds;

(iv)   water-jacket rather than immersion heater type temperature control (to avoid localised heating at low stirring speeds);

(v)   top driven stirrer so that cells cannot become entangled between moving parts.

As long as adequate mixing, and thus mass transfer of oxygen into the vessel can be maintained without damaging the cells, there is no maximum to the scale-up potential. BHK21 cells have been grown in 2000 l vessels for the production of foot and mouth disease vaccine. There are many heteroploid cell lines, such as Vero and HeLa, which would grow in such systems. However, while regulatory agencies require pharmaceutical products to be manufactured predominantly from normal diploid cells (which will not grow in suspension), the only stimulation to use this scale of culture is for veterinary products. Meanwhile, there are many applications for research products from various types of cells, and this is served by the 2 — 50 l range of vessels. At present, the greatest incentive for large scale systems is to grow hybridoma (monoclonal antibody producing) cells. In tissue culture the antibody yield is 50- to 100-fold lower than when the cells are passaged through the peritoneal cavity of mice. At the moment, the need is to supply antibodies to meet the requirements for diagnostic and affinity chromatography purposes. The future need will be to supply them for therapeutic and prophylactic drug treatments. Many of these cells are fragile and low yielding in culture, and the specialised techniques and apparatus described in Section 4.7 are partly aimed towards this type of cell.

4.6 Continuous Flow Culture

4.6.1 *Introduction*

At submaximal growth rates, the growth of a cell is determined by the concentration of a single growth-limiting nutrient. This is the basis of the chemostat, a fixed volume culture, in which medium is fed in at a constant rate, mixed with the cells, and then leaves at the same rate. The culture begins as a batch culture while the inoculum grows to the maximum value that can be supported by the growth-limiting nutrient (assuming the dilution rate is less than the maximum growth rate). As the growth-limiting nutrient decreases in concentration, so the growth rate declines until it equals the dilution rate. When this occurs, the culture is defined as being in a 'steady-state' as both the cell numbers and nutrient concentrations remain constant.

When the culture is in a steady-state, the cell growth rate (μ) is equal to the dilution rate (D). The dilution rate is the quotient of the medium flow rate (f) per unit time and the culture volume (V):

$$\mu = D = f/V \ \text{day}^{-1} \qquad \text{Equation 8}$$

As the growth rate is dependent on the medium flow rate, the mean generation (doubling) time can be calculated:

$$t_d = \ln2/D \qquad \text{Equation 9}$$

An alternative system to the chemostat is the turbidostat in which the cell density is held at a fixed value by altering the medium supply. The cell density (turbidity) is usually measured through a photo-electric cell. When the value is below the fixed point, medium supply is stopped to allow the cells to increase in number. Above the fixed point, medium is supplied to wash out the excess cells. This system only really works well when the cell growth rate is near maximum. However, this in fact is its main advantage over the chemostat which is least efficient, or controllable, when operating at the cell's maximum growth rate. Continuous flow culture of animal cells has been well reviewed by Tovey (21).

4.6.2 *Equipment*

Complete chemostat systems can be purchased from all dealers in fermentation equipment. However, systems can be easily constructed in the laboratory (*Figure 13*) (22). The culture vessel needs a side arm overflow at the required liquid level, which should be approximately half the volume of the vessel. If a suitable 37°C cabinet is not available, then a water-jacketed vessel is needed. Apart from this, all other components are standard laboratory items. Vessel enclosures can be made from silicone (or white rubber) bungs wired onto the culture vessel. A good quality peristaltic pump, such as the Watson-Marlow range, is recommended.

4.6.3 *Experimental*

Recommended cells are LS, HeLa-S3 or an established lymphoblastic cell line, such as L1210. Growth-limiting factors can be chosen from the amino acids or glucose.

(i)   Inoculate the culture vessel at 10^5 cells/ml in preferred medium (e.g., Eagle's MEM with 10% calf serum and the growth-limiting factor).

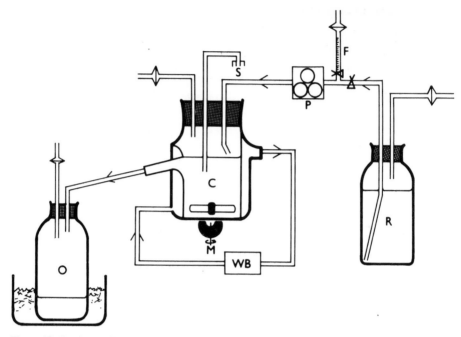

Figure 13. Continuous-flow culture system (22). C, water-jacketed culture vessel; WB, water bath and circulation system; O, overflow vessel; R, reservoir; P, pump; S, sampling device; F, burrette for measuring flow rate; M, magnetic drive.

(ii)   The chosen dilution rate is turned on after 24 − 48 h of growth. The maximum rate must not exceed the maximum growth of the cell line, which is usually within the range 14 − 27 h doubling time (although some cells can double their number in 9 h). Thus, the dilution rate will be in the range of 0.1 day^{-2} (td = 166 h) to 1.2 (td = 14 h).

(iii)   A steady-state will become established within 100 − 200 h, although this may take up to 400 h, especially at the low dilution rates. This will be recognised by the fact that the daily cell counts will not vary by more than the expected counting error.

(iv)   The culture can be maintained almost indefinitely, assuming it is kept sterile and no breakdown in components occurs. Sometimes the culture has to be terminated because of excessive attachment of cells at the interfaces. A duration of 1000 h is considered satisfactory.

(v)   Once the steady-state has been demonstrably achieved, the flow rate can be altered and the response of the cell population in establishing new steady-state conditions can be studied.

(vi)   As well as carrying out routine cell counts, measurements can also be made to demonstrate the homogeneity of the cells and medium. For instance, the size and chemical composition of the cells is remarkably consistent. Also, some of the medium components should be measured (e.g., glucose, an amino acid, lactic acid) to demonstrate again the consistency of the culture over a long period of time.

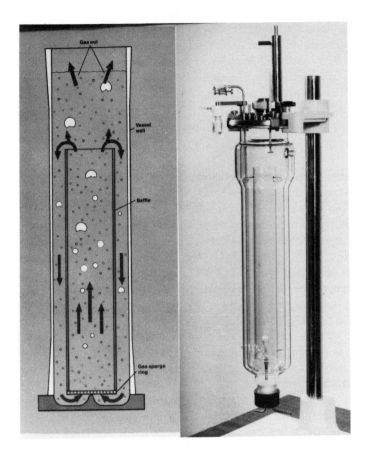

Figure 14. Principle of the airlift fermenter.

4.6.4 *Uses*

(a)   A readily available continuous source of cells.

(b)   As optimal conditions or any physiological environment, can be maintained, the culture is very suitable for product generation (as already shown for viruses and interferon) (21). For many purposes a 2-stage chemostat is required so that optimal conditions can be met for cell growth (first stage) and product generation (second stage).

4.7 **Airlift Fermenter**

The airlift fermenter relies on the bubble column principle to both agitate and aerate a culture. Instead of mechanically stirring the cells, air bubbles are introduced into the bottom of the culture vessel. An inner (draft) tube is placed inside the vessel and mixing occurs because the air bubbles lift the medium (aerated medium has a lower density than non-aerated medium). The medium and cells which spill out from the top of the draft tube then circulate down the outside of the vessel. The amount of energy (compressed air) needed for the system is very low, shear forces are absent, and this method is thus ideal for fragile animal and plant cells. Also as oxygen is continuously supplied to the culture, the large number of bubbles results in a high mass transfer rate. Culture units, as illustrated in *Figure 14*, are commercially available in sizes from 2 to 90 l (LH Fermentation). The one disadvantage of the system is that scale-up is more or less linear (the 90 l vessel requires nearly 4 m headroom). Whether it will be possible to use multiple draft tubes, and thus enable units with greatly increased diameters to be used, remains a developmental challenge.

4.8 Encapsulated Cells

The entrapment of cells in semi-solid matrices, or spheres, has many applications but the basic function is to stabilise the cell and thus protect it from sub-optimal conditions. Cells can be immobilised by adsorption, covalent bonding, cross-linking or by entrapment in a polymeric matrix. Materials that can be used are gelatin, polylysine, alginate and agarose; the choice largely depends upon the problem being addressed. These techniques have been used to protect cells that are being transported or posted between laboratories; to store cells at 4°C for an extended period of time (e.g., 5 weeks); to avoid immune rejection of transplanted cells; and to protect fragile cells (e.g., hybridomas) from mechanical stress in large-scale culture equipment (23,24). The latter application not only makes it possible to use such equipment but, also, allows production of hormones, antibodies, immunochemicals and enzymes over much longer periods than is possible in homogeneous suspension culture. The matrix allows free diffusion of nutrients and generated product between the enclosed microenvironment and the external medium.

Alginate is a polysaccharide and is cross-linked with Ca^{2+} ions. The rate of cross-linking is dependent on the concentration of Ca^{2+} (e.g., ~30 min with 10 mM $CaCl_2$). A recommended technique is to suspend the cells in isotonic NaCl buffered with Tris (1 mM) and 4% sodium-alginate, and to add this mixture dropwise into a stirred solution of isotonic NaCl, 1 mM Tris, 10 mM $CaCl_2$ at pH 7.4. The resulting spheres are 2 – 3 mm in diameter. The entrapped cells can be harvested by dissolving the polymer in 0.1 M EDTA or 35 mM sodium citrate. Disadvantages of alginate are that calcium must be present and phosphate absent and that large molecules, such as monoclonal antibodies, cannot diffuse out. For these reasons, agarose in a suspension of paraffin oil provides a more suitable alternative. 5% agarose in Ca^{2+} and Mg^{2+} free PBS is melted at 70°C, cooled to 40°C, and mixed with cells suspended in their normal growth medium. This mixture is added to an equal volume of paraffin oil and emulsified with a Vibro-mixer. The emulsion is cooled in an ice-bath, growth medium added and, after centrifugation, the oil is removed. The spheres (80 – 200 μm) are washed in medium, centrifuged and, after removing the remaining oil, transferred to the culture vessel.

The use of entrapped cells in fermenter culture is an attractive technique for product generation. It has many of the advantages of microcarriers (spheres with cell growth on the outer surface) in that medium changes, and alterations in the cell to medium volume ratio, can be performed easily. In addition, entrapment can be used to facilitate perfusion, or medium changes, and products can be harvested cell-free.

5. REFERENCES

1. Griffiths,J.B. (1972) *J. Cell Sci.*, **10**, 512.
2. Good,N.E. (1966) *Biochemistry*, **5**, 467.
3. Leibovitz,A. (1963) *Am. J. Hyg.*, **78**, 173.
4. Spier,R.E. and Griffiths,J.B. (1984) *Dev. Biol. Stand.*, **55**, 81.
5. Toth,G.M. (1977) in *Cell Culture and its Applications*, Acton,R.T. and Lynn,J.D. (ed.), Academic Press, NY, p. 617.
6. Griffiths,J.B. (1984) *Dev. Biol. Stand.*, **55**, 113.
7. Maroudas,N.G. (1975) *J. Theor. Biol.*, **49**, 417.
8. Kruse,P.J., Keen,L.N. and Whittle,W.L. (1970) *In Vitro*, **6**, 75.
9. Corbiel,M., Trudel,M. and Payment,P. (1979) *J. Clin. Microbiol.*, **10**, 91.
10. Skoda,R., Pakos,V., Hormann,A., Spath,O. and Johansson,A. (1979) *Dev. Biol. Stand.*, **42**, 121.
11. Munder,P.G., Modolell,M. and Wallach,D.F.H. (1971) *FEBS Lett.*, **15**, 191.
12. Jensen,M.D. (1981) *Biotech. Bioeng.*, **23**, 2703.
13. Burbidge,C. and Darcey,I.K. (1984) *Dev. Biol. Stand.*, **55**, 255.
14. Whiteside,J.P. and Spier,R.E. (1981) *Biotech. Bioeng.*, **23**, 551.
15. Weiss,R.E. and Schleiter,J.B. (1968) *Biotech. Bioeng.*, **10**, 601.
16. van Wezel,A.L. (1967) *Nature*, **216**, 64.
17. Griffiths,J.B. and Thornton,B. (1982) *J. Chem. Technol. Biotechnol.*, **32**, 324.
18. Spier,R.E. and Whiteside,J.P. (1984) *Dev. Biol. Stand.*, **55**, 151.
19. Capstick,P.B., Garland,A.J., Masters,R.C. and Chapman,W.G. (1966) *Exp. Cell Res.*, **44**, 119.
20. Feder,J. and Tolbert,W.R. (1983) *Sci. Am.*, **248**, 24.
21. Tovey,M.G. (1980) *Adv. Cancer Res.*, **33**, 1.
22. Pirt,S.J. and Callow,D.S. (1964) *Exp. Cell Res.*, **33**, 413.
23. Nilsson,K. and Mosbach,K. (1980) *FEBS Lett.*, **118**, 145.
24. Nilsson,K., Scheirer,W., Merton,O.W., Ostberg,L., Liehl,E., Katinger,H.W.D. and Mosbach,K. (1983) *Nature*, **302**, 629.

6. APPENDIX

Companies Supplying Fermentation Equipment

Bioengineering AG., Ch-8636 Wald/ZH, Switzerland.
Biolafitte, Poissy, France.
B. Braun AG., Melsungen, FRG.
Chemap AG., Mannedorf, Switzerland.
Giovanola Freres, SA, Montley, Switzerland.
LH Fermentation, Stoke Poges, UK.
New Brunswick Scientific Co., Inc., NJ, USA.
Setric GI B.P., Toulouse, France.

Other Companies Quoted in Text

APV Co. Ltd., Crawley, Sussex, UK.
Amicon Corporation, Danvers, MA, USA.
Becton Dickinson (UK) Ltd., Oxford, UK.
Bellco Glass Inc., Vineland, NJ, USA.
Corning Ltd (UK), Stone, UK.
Dr. W. Ingold AG., Zurich, Switzerland.
Flow Laboratories Ltd., Irvine, UK.
Gallenkamp & Co. Ltd., London EC2P 2ER, UK.
K. C. Biological Inc., Kansas, USA.
Microflow Ltd., Fleet, Hants, UK.
Nunc A/S, Roskilde, Denmark.
Pall Filtration Ltd., Portsmouth, UK.
Pharmacia Fine Chemicals, Uppsala, Sweden.
Russel pH Ltd., Fife, UK.
Sterilin Ltd., Teddington, UK.
Techne (Cambridge) Ltd., Duxford, UK.
Watson Marlow, Falmouth, UK.
Wheaton Scientific, Millville, NJ, USA.

CHAPTER 4

Preservation and Characterisation

R.J.HAY

1. INTRODUCTION

Literally thousands of different cell lines have been derived from human and other meta-zoan tissues[1]. Many of these originate from normal tissues and exhibit a definable, limited doubling potential. Other cell lines may be propagated continuously, either having gone through a change from the normal primary population or having been developed initially from tumor tissue. Both finite lines of sufficient doubling potential and continuous lines can be expanded to produce a large number of aliquots, frozen and characterised for widespread use in research.

The advantages of working with a well-defined cell line free from contaminating organisms would appear obvious. Unfortunately, however, the potential pitfalls associated with the use of cell lines obtained and processed casually require repeated emphasis. Numerous occasions where cell lines exchanged among cooperating laboratories have been contaminated with cells of other species have been detailed and documented elsewhere (2,3). For example, lines supposed to be human have been found to be monkey, mouse or mongoose while others thought to be monkey or mink were identified as rat and dog (2). Similarly, the problem of intraspecies cross-contamination among cultured human cell lines has been recognised for over 20 years and detailed reviews are available on the subject (4). The loss of time and research funds as a result of these problems is incalculable.

While bacterial and fungal contaminations represent an added concern, in most instances they are overt and easily detected, and are therefore of less serious conseqeunce than the more insidious contaminations by mycoplasma. That the presence of these micro-organisms in cultured cell lines often negates research findings entirely has been stated repeatedly over the years by a number of specialists in mycoplasma biology (5,6). However, the difficulties of detection and prevalence of contaminated cultures in the research community suggest that the problem cannot be over-emphasised.

In this chapter the procedures used to preserve and characterise cell lines and hybridomas at the American Type Culture Collection (ATCC) are outlined. They have evolved over the past 20 years as awareness of the difficulties of microbial and cellular cross-contaminations became apparent and as the need for well-characterised cell lines increased.

---

[1][Use of the terms cell line and cell strain is as recommended by The Tissue Culture Association committee on nomenclature (1)].

2. CELL LINE BANKING

Cell lines pertinent for accessioning are selected by ATCC scientists and advisors during regular reviews of the literature. The originators themselves also frequently offer the lines directly for consideration. Detailed information with regard to specific groups of cell lines accessioned is available elsewhere (7−9). Generally, starter cultures or ampoules are obtained from the donor, and progeny are propagated according to instructions to yield the first 'token' freeze. Cultures derived from such token material are then subjected to critical characterisations as described below. If these tests suggest that further efforts are warranted the material is expanded to produce seed and distribution stocks. Note especially that the major characterisation efforts are applied to cell populations in the initial seed stock of ampoules. The distribution stock consists of ampoules that are distributed on request to investigators. The reference seed stock, on the other hand, is retained to generate further distribution stocks as the initial stock becomes depleted.

Although this procedure has been developed to suit the needs of a large central repository, it is also applicable in smaller laboratories. Even where the number of cell lines and users may be limited, it is important to separate 'seed stock' from 'working or distribution stock'. Otherwise the frequent replacement of cultured material, recommended to prevent phenotypic drift or senescence, may deplete valuable seed stock which may be difficult and expensive to replace.

These various steps in the overall accessioning scheme are summarised in *Figure 1*. It is important to recognise that the characterised seed stock serves as a frozen 'reservoir' for production of distribution stocks over the years. Because seed stock ampoules are used to generate new distribution material, one can ensure recipients that all the cultures obtained closely resemble those received 2, 5, 10 or more years previously. *This is a most critical consideration for design of cell banking procedures.* Problems

Accessioning Scheme

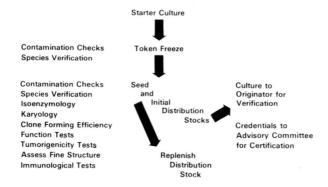

Figure 1. Scheme illustrating a recommended plan for accessioning cell lines to be banked for general distribution. At least the first two of the various characterisations indicated on the left hand side should be performed before release of any cell line.

associated with genetic instability, cell line selection, senescence or transformation may be minimised or avoided entirely by strict adherence to this principle.

It is prudent to handle all primary tissues or cell lines not specifically shown to be free of adventitious agents as biohazards in a Class II vertical laminar flow hood (BS5726; NIH Spec. 03-112). This precaution protects both the cell culture technician and the laboratory from infection or contamination. Some recommend furthermore, that all human tumour lines be treated with similar caution due to the known presence of on-cogenes and the recent demonstration of transfection at least within cell culture systems.

3. CELL FREEZING AND QUANTITATION OF RECOVERY

Cellular damage induced by freezing and thawing is generally believed to be caused by intracellular ice crystals and osmotic effects. The addition of a cryoprotective agent, such as dimethyl sulphoxide (DMSO) or glycerol, and the selection of suitable freezing and thawing rates minimises cellular injury.

While short-term preservation of cell lines using mechanical freezers ($-75°C$) is possible, storage in liquid nitrogen ($-196°C$) or its vapour (to $-120°C$) is much preferred. The use of liquid nitrogen refrigerators is advantageous not only because of the lower temperatures and, consequently, almost infinite storage times possible, but also due to the total absence of risk of mechanical failures and the prolonged holding times now available. Certainly for all but the smallest cell line banking activity, storage in a liquid nitrogen refrigerator is essential.

Two considerations on safety in processing cell lines require special emphasis. Firstly, the cell culture technician may be endangered due to the possibility that liquid nitrogen can penetrate ampoules *via* hairline leaks during storage. On warming, rapid evaporation of the nitrogen within the confines of such an ampoule can cause a sharp explosion with shattered glass flying at high force almost instantaneously in all directions. Fortunately the frequency of such traumatic accidents declines dramatically as the operator gains experience in ampoule sealing and testing (Sections 3.1.1 and 3.3). However, even with highly accomplished laboratory workers, the remote possibility of explosion still exists. Thus, a protective face mask should be worn whenever ampoules are removed from liquid nitrogen storage.

Secondly, DMSO can solubilise organic substances and, by virtue of its penetrability through rubber and skin, carry these to the circulation. Thus, special precautions should be exercised when using DMSO to avoid contamination with hazardous chemicals and minimise skin contact.

3.1 Equipment

3.1.1 *Ampoules, Marking and Sealing Devices*

A decision on whether to utilise glass or plastic ampoules will depend on the scale of operation and the extent of anticipated distribution. Glass ampoules can be sterilised, loaded with the appropriate cell suspension and permanently sealed in comparatively large quantities. They are recommended for large lots of cells being prepared for long-term use or general distribution. Smaller numbers of plastic ampoules are easier than glass to handle, mark and seal. Problems with the seal may occur in some cases, however, especially if frequent handling or manipulations for shipment are necessary.

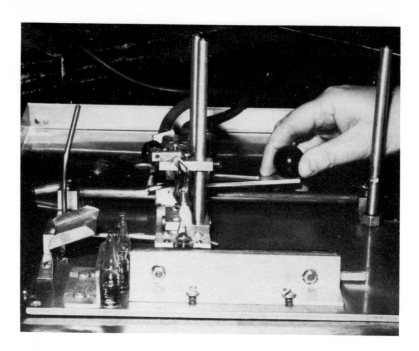

Figure 2. Device for heat-pull sealing of glass ampoules. The ampoules move from right to left and are automatically rotated as each tip is heated in the flame. At maximum temperature the tip is pulled upward as shown to ultimately form a seal of molten glass (left hand side). The speed at which the handle is moved is critical for production of an appropriate seal.

The marking of ampoules requires special consideration in that legibility can easily be obscured as the ampoules are frozen, snapped on and off storage canes, transferred between freezers or shipping containers and so forth. The use of paper labels, ball-point pen or standard laboratory markers all are problematical and this is especially true with glass ampoules. These should be labelled in advance with ceramic ink that can be heat annealed to the glass surface. The markings can be applied by hand with a straight pen or, if large lots are being processed, through use of a mechanical labeller (e.g., from Markem Co., Keene, New Hampshire).

Glass ampoules can best be sealed by pulling on the neck of the ampoule as it is rotated in the highest heat zone of the flame from a gas-oxygen torch. This pull-seal technique is preferred since it reduces the risk of permitting pinhole leaks in the sealed tip. Torches for manual sealing can be obtained from scientific supply houses. For large lots, a torch can be attached to a semi-automatic sealing device (*Figure 2*) available as Bench sealer model 161 from Morgan Sheet Metal Co., Sarasota, FL, USA.

3.1.2 *Slow Freezing Apparatus*

The optimum freezing rate for cell lines (usually about $-1\,°C/min$) can be achieved through use of apparatus varying in complexity from a tailor-made styrofoam box to a completely programmable freezing unit. The former should have a wall thickness of about 2 cm to approximate the $1\,°C$ cooling rate when placed in a mechanical freezer at −70°C. Alternatively, manufacturers of liquid nitrogen refrigerators supply adapted refrigerator neck plugs, at modest cost, which can be adjusted for slow freezing of small quantities (1−9) of ampoules. For those who produce larger quantities of ampoules and require more precise control of the freezing rate, freezing units which have constant rate or programmable, step-wise freezing rates (Cryo-Med; Planer; Union Carbide) should be considered.

3.1.3 *Liquid Nitrogen Refrigerators*

Choice of an appropriate refrigerator will require considerations of economy, both in terms of liquid nitrogen consumption and initial outlay, storage capacity required and desired ease of entry and retrieval. Freezers with narrower neck tube openings are generally more economical. Some larger models are equipped with lazy susan trays for convenient operation.

Ampoules of cells may be stored immersed in liquid nitrogen or in the vapour phase. The latter has the advantage that ampoules with pinhole leaks will not be exposed to liquid so the danger of explosion is eliminated. The slightly higher temperature probably offers no disadvantage except perhaps with seed stocks retained for extremely long periods.

Useful accessories for refrigerators include roller bases for ease of movement, alarm systems to warn of dangerously-low levels of liquid nitrogen and racking systems, with larger refrigerators, for ready storage and recovery. Even when an automatic alarm system is used this should be backed up with a regular manual check using a dip-stick. Electronic systems can fail and it has been known for both the automatic fill system and its back-up alarm to fail.

3.2 **Preparation and Freezing**

Cultures in the late logarithmic or just pre-confluent phase of growth should be selected to give the highest possible initial viability. Treat the cultures with trypsin if necessary to produce a uniform single cell suspension as if for routine subcultivation and proceed as follows.

(i)   Just before use, prepare the freeze medium by simple admixture of fresh growth medium and the required cryoprotective agent. The concentrations of choice vary slightly with the cell line and cryoprotective agent used, ranging from 5 to 10% (v/v).

(ii)   Collect the cells in pellet form by centrifugation at about 200 g for 10 min and resuspend in an appropriate volume of freeze medium at room temperature. Concentrations of $10^6 - 10^7$ cells/ml are generally satisfactory and practical. For some applications it may be desirable to increase or decrease this by as much as 1 log.

(iii)   If the total volume is large, as for a production-level freeze, maintain the cell suspension with gentle agitation in an appropriate stirring vessel. Set this up in advance complete with a Cornwall automatic syringe. If the cell suspension is in a small volume (10−20 ml), dispense by means of a syringe fitted with an 18-gauge needle or cannula. Mix repeatedly during this process.

(iv)   Maintain the pH when necessary by gassing with an appropriate mixture of water-saturated air/CO_2.

(v)   Dispense the cell suspension in 1-ml aliquots to the ampoule taking care to proceed rapidly and with uniformity. Cells sediment quickly under unit gravity in the Cornwall tubing and inaccuracies can result if the process is interrupted or arhythmic.

(vi)   Seal each ampoule, place on an ampoule cane or in a suitable rack and immerse totally upright in aqueous methylene blue (0.05%) at 4°C. After 30−45 min remove, wash in cold tap water and discard any ampoule containing the blue dye.

(vii)   Dry the ampoules thoroughly and begin slow freezing. The optimum rate of cooling is usually 1°C/min but varies among different cell lines and can best be determined empirically before any large-scale freeze is attempted.

(viii) When ampoules have reached −50 to −70°C or lower, remove and transfer them immediately and rapidly to the liquid nitrogen refrigerator.

(ix)   Record the location and specifics of the cell line in question. Two separate cross-index files providing both subject and location cards can be used for this purpose.

3.3 Reconstitution and Quantitating Recovery

Rapid thawing of the cell suspension is essential for optimal recovery.

(i)   With face and neck protected by a full face mask, retrieve the selected ampoule rapidly and plunge directly into a warm water bath (37°C). If ampoules have been stored submerged in liquid nitrogen each ampoule must be thawed separately by depositing directly into about 10 cm of water at 37°C in a 1.5−2 litre bucket with a lid, snapping the lid closed immediately after the ampoule is inserted.

(ii)   Agitate the ampoule contents until the suspension has thawed completely (20−60 sec).

(iii)   Immerse the ampoule in 70% ethanol at room temperature.

(iv)   Score standard ampoules using a small file that has been dipped previously in ethanol. Pre-scored ampoules require no such treatment.

(v)   Use a sterile towel to pick up the ampoule; break sharply at the neck at the pre-scored point.

(vi)   Transfer the contents, by means of a sterile Pasteur pipette or syringe, to a centrifuge tube or culture vessel. Add 10 ml of complete growth medium. In some cases, slow addition over a 1−2 min interval at this point may be beneficial.

(vii)   Centrifuge at 200 g for 10 min. Remove the supernatant, add a fresh aliquot of medium, and mix the suspension. This centrifugation step may be omitted in some cases since the residual concentration of cryoprotective agent is low and the stress of centrifugation can be harmful.

(viii) Initiate the culture by standard procedure. If the centrifugation step was omitted, change the medium after 24 h.

A variety of methods may be used to quantitate cell recovery after freezing. Of course this is the first characterisation step to be performed after preservation.

3.3.1 Dye Exclusion for Quantitating Cell Viability

A very approximate estimate of the viability of cells in a suspension may be obtained by the dye exclusion test. A solution of the dye in saline is added to the suspension and the percentage of cells which do not take up the stain is determined by direct count using a haemocytometer. Trypan blue or erythrocin B are probably the stains most commonly used for this procedure. The former reportedly has a higher affinity for protein in solution than for non-viable cells. This may reduce the accuracy of the estimate if the suspension contains much more than 1% serum. Furthermore, because solutions of erythrocin B are clear, microbial growth or precipitates are immediately apparent. This is not true for stock solutions of trypan blue. A recommended procedure can be outlined as follows.

(i) Obtain a uniform suspension of cells in growth medium by any standard procedure. If the estimate is for freeze-characterisation, reconstitution of contents from about 5% of the ampoules prepared is recommended.

(ii) Dilute an accurately measured aliquot of the cell suspension using the stock dye solution. This consists of 100 mg erythrocin B/100 ml of an isotonic phosphate-buffered saline (PBS) adjusted to pH $7.2-7.4$ with 1 M NaOH. For ease and accuracy of count a final density of about $0.3-2 \times 10^6$ cells/ml is satisfacotry.

(iii) Score the number of stained *versus* total number of cells keeping in mind that, in general, a larger sampling will give a more accurate quantitative estimate. Ideally the count should be peformed within $1-5$ min after mixing of the dye and cell suspension.

3.3.2 *Clone-forming Efficiency*

The dye-exclusion test for cell viability generally overestimates recovery. For lines consisting of adherent cells, the clone-forming ability of cells from the reconstituted population represents a more accurate overall estimate of survival. Of course the choice of growth medium used, the substrate on which the clones develop, the incubation time and so forth all may also have an effect on the end result. Thus for comparisons among different freezes, conditions for selected lines must be standardised. A representative outline as used at the ATCC is provided below.

(i) Pool the reconstituted contents of ampoules from about 5% of a freeze lot and serially dilute the suspension to provide inocula of $100-10^4$ cells/culture depending upon the cell line, plate size and expected recoveries. For example, cells of a line with high plating efficiency could be added at 10 cells, 100 cells, 1000 cells and 10 000 cells to four separate plate sets, assuming 9 cm plates were used.

(ii) Make the serial dilutions by 10-fold reductions in cell number by transferring 1 ml of suspension to 9 ml of the selected, complete growth medium. Half-log dilutions can be made by transferring 1.0 ml of cell suspension to 2.16 ml of diluent medium. Discard the 1 ml pipette and use a fresh 1 ml pipette to mix and make each subsequent, lower density transfer for dilution.

(iii) Inoculate 1 ml of each dilution, beginning with the lowest cell concentration, to each of three T-75 flasks (or other suitable culture vessel) containing 8 ml of growth medium and incubate at 37°C.

(iv) Renew the fluid on test cultures on the fourth or fifth day and thereafter every third or fourth day.

(v) After $12-14$ days total, remove the fluid, and fix the culture with a 10% solution of formaldehyde or other suitable fixative.

(vi)   Remove the fixative, rinse the flask interior gently with several changes of tap water, and stain with 1 % aqueous toluidine blue or other simple stain for 1 − 5 min.

(vii)   Remove the staining solution, rinse out residual fluid with tap water and count clones consisting of 16 or more cells through use of a dissecting microscope.

(viii) Calculate the percent clone-forming efficiency:

$$\frac{\text{clones formed}}{\text{number of cells inoculated}} \times 100$$

3.3.3 *Proliferation in Mass Culture*

The vitality of reconstituted cells from either adherent or non-adherent lines can also be documented by simply quantitating proliferation during the initial 1 − 2 weeks after recovery from liquid nitrogen as follows.

(i)   Using the cell suspension pooled from 5 % of a freeze lot, inoculate three sets of culture vessels (e.g., T-25 flasks) at different densities. Typically one might choose the inoculum expected to yield a confluent or maximum density culture in 7 days for one set; twice that for a second set and 3 − 5 times that (depending on viable count) for the third set.

(ii)   Renew the culture fluid (or add fresh medium for a suspension culture) after 4 days.

(iii)   Harvest the three sets of cultures by standard trypsinisation if an adherent culture or by direct sampling if a suspension culture, and determine the cell yield by either electronic cell count or use of a haemocytometer.

(iv)   Calculate the fold increase (no. cells recovered/no. cells inoculated) and compare this with what one would expect under similar conditions with cells taken during logarithmic or pre-confluent growth phase.

Typical results from the above tests have been recorded for a wide variety of cell lines in the ATCC Catalogue of Strains II (8).

4. CELL LINE CHARACTERISATIONS

In addition to recoverability from liquid nitrogen, the absolute minimum recommended characterisations include verification of species and demonstration that the cell line is free of bacterial, fungal or mycoplasmal contamination.

4.1 **Species Verification**

Species of origin can be determined for cell lines by a variety of immunological tests by isoenzymology and/or by cytogenetics (10). Advantages and problems associated with each of these will be included with the appropriate sections.

4.1.1 *Fluorescent Antibody Staining*

The indirect fluorescent antibody staining technique is used at the ATCC as one method for verifying the species of cell lines. The technique involves two general steps. Firstly, species-specific antiserum, produced in rabbits (*Table I*), is used to label test cells plus positive and negative control cell populations. Secondly, anti-rabbit globulin, pro-

Table 1. Preparation of Antiserum to Cultured Cells.

1. Harvest cells of known species by scraping from the culture surface with a rubber policeman.
2. Wash by suspending in Hank's balanced salt solution (HBSS)[a] with subsequent centrifugation at 200 g for 10 min and repeat three times.
3. Resuspend in HBSS such that the viable cell count is $\sim 5 \times 10^5$/ml (first week), 10^6/ml (second week) and 10^7/ml (third week).
4. Inoculate 1 ml to each marginal ear vein of a healthy rabbit twice weekly for 3 weeks, increasing the dose each week, as in step 3.
5. Administer three or more additional booster injections at 10^7 cells/ml (1 ml/ear) on a weekly basis.
6. After the third booster injection, perform test bleedings, collect and serially dilute antisera. Mix an equal volume of cell suspension (10^6/ml) and evaluate cytotoxicity using the viable staining technique described earlier.
7. If the titres are satisfactory (1:8 or greater) collect the blood by cardiac puncture, permit it to clot for $1-2$ h at room temperature and centrifuge at 200 g for 15 min.
8. Remove serum, inactivate complement by heating at 56°C for 30 min, dilute and distribute in 0.2 ml aliquots for storage at -70°C.

[a]All solutions are prepared in double glass-distilled water (DGDW) unless otherwise indicated.
See reference (10) for more detail.

duced in goats and coupled to the fluorescent dye, fluorescein isothiocyanate (FITC) is applied. The second reagent binds to the rabbit antibody which has attached to the target cells and, by virtue of the fluorescence, the antibody-antigen complexes can be visualised.

To verify species on a test cell line:

(i)   Harvest by trypsinisation if necessary and wash three times by suspending the cells in PBS at pH 7.5 with subsequent centrifugation to form a cell pellet.
(ii)   Resuspend the washed cells in PBS to give a density of $3-4 \times 10^6$/ml.
(iii)   Mix 0.1 ml of cell suspension and 0.1 ml of diluted antiserum and place in a humidified incubation chamber at room temperature for 30 min. The appropriate dilution of antisera will have to be determined empirically for each antiserum preparation with positive control cells through an initial preliminary trial. Non-specific absorption can usually be excluded by further diluting the antiserum.
(iv)   Samples are then washed to remove unadsorbed antiserum using three complete changes of PBS and are incubated for 30 min in the dark with 0.1 ml of FITC-conjugated, goat anti-rabbit antiserum (obtainable commercially).
(v)   After a final three additional washes with PBS, a drop of the final cell suspension is sealed under a coverslip. This is examined by fluorescence microscopy at 500× using number 50 barrier and BG12 exciter filters on a Zeiss Universal microscope with an epi-illuminator.
(vi)   Positive reactions are seen as staining of brilliant fluorescence intensity at the cell periphery (*Figure 3*).
(vii)   Controls consisting of cells of the suspected species, one related species and a distant species are included with each test.

This method is an adaptation of that described by Stulberg (10). Advantages over other methods include simplicity and the ability to allow identification of even minor cellular contaminants among populations. Stulberg reported that as few as one con-

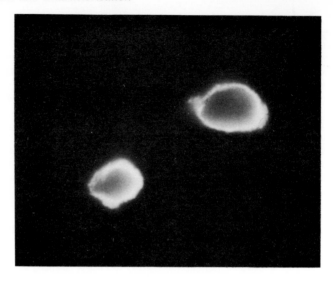

Figure 3. Indirect fluorescent antibody test for cell species. Note the halo of fluorescence around the cells showing a typical positive reaction. (Photograph courtesy of M. L. Macy.)

Table 2. Preparation of Cell Extract.

1. Obtain a cell suspension by standard methods and wash three times in 0.9% NaCl solution (pH 7.1) containing EDTA at 6.6×10^{-4} M by centrifugation with subsequent resuspension as usual.
2. Resuspend the resulting cell pellet at $2-5 \times 10^7$ cells/ml in saline-EDTA.
3. Produce an homogenate by freezing and thawing rapidly three times in liquid nitrogen or a dry ice-alcohol bath, by ultrasonic treatment or by mixing with an equal volume of octylalcohol (4°C overnight).
4. Dispense aliquots of the homogenate to Eppendorf tubes, clarify by centrifugation in a microfuge for $1-2$ min and store at -70°C for subseqeunt assay.

taminating cell among 10 000 could be identified under appropriate conditions of resolution.

4.1.2 *Isoenzyme Profiles*

Isozyme analyses performed on homogenates of cell lines from over 25 species have clearly demonstrated the utility of this biochemical characteristic for species verification (11). By determining the mobilities of three isozyme systems [glucose-6-phosphate dehydrogenase (G6PD), lactic dehydrogenase (LDH), and nucleoside phosphorylase (NP)] using vertical starch gel electrophoresis, one can identify species of origin of cell lines with a high degree of certainty. The procedures are relatively straightforward, provide consistent results and do not require expensive equipment.

A method for preparing cell extracts is summarised in *Table 2*. The compositions of stock buffers for starch electrophoresis are given in *Table 3*. Ingredients for electrophoretic separations and staining for the three enzyme systems are provided in *Table 4*.

Prepare the starch gel as follows.

(i)   Suspend 60 g of electrostarch in 500 ml of the appropriate buffer and produce a uniform suspension by stirring and heating to 90°C. This may be accomplish-

Table 3. Stock Buffers for Electrophoresis.

Tris-citrate (TC)	1 × Stock
Tris	17 g
Citric acid H$_2$O	9.1 g
Distilled H$_2$O	1000 ml
pH	7.1

Tris-EDTA-Borate (TEB)	5 × Stock
Tris	109.0 g
EDTA Na$_2$	7.6 g
Boric acid	6.2 g
Distilled H$_2$O	1000 ml
pH	8.6

Gel buffers: 50 ml of 1 x stock + 450 ml H$_2$O
Tray buffers: 1 x stock solution

Table 4. Procedures for LDH, G6PD and NP[a].

Enzyme system	Chamber buffer	Starch buffer	Staining solutions[b]
LDH	TC	TC (0.1×)	10 ml H$_2$O 5 ml 0.5 M Tris pH 7.5 5 ml 1.0 M Na lactate 5 ml NAD[c] (10 mg/ml) 5 mg MTT[c] 2 mg PMS[c]
G6PD	TEB	TEB (0.1×)	5 ml H$_2$O 5 ml 0.5 M Tris pH 7.5 5 ml 0.025 M glucose-6-phosphate 5 ml 0.1 M MgCl$_2$.6H$_2$O 5 ml 0.005 M NADP[c] 5 mg MTT[c] 2 mg PMS[c]
NP	TEB	TEB (0.2×)	20 ml H$_2$O 5 ml 0.1 M NaH$_2$PO$_4$.H$_2$O 50 ml inosine 100 μl xanthine oxidase[d] 5 mg MTT[c] 2 mg PMS[c]

[a]All gels are run for 16 – 18 h at 4°C at 4 – 5 V/cm (160 – 180 V). NADP (5 ml at 0.005 M) is added to the cathode buffer as well as to the staining solution for G6PD.
[b]These stock solutions are mixed just before use with an equal volume (25 ml) of 2% noble agar in water which has been liquified and cooled to 45 – 50°C.
[c]NAD, nicotinamide adenine dinucleotide; NADP, NAD-phosphate; MTT, dimethylthiazol diphenyl-tetrazolium; PMS, phenazine methosulphate.
[d]At 100 mg/ml - add just before mixing with agar.

ed using a boiling water bath, a stirrer-hot plate, a vacuum or microwave oven. The stirring must be constant to avoid lumpiness and burning.

(ii) Apply a vacuum for about 60 sec to remove gas bubbles, and allow the suspension to cool uniformly to 60°C by swirling gently.

(iii)   For LDH or G6PD systems add NADP at this point (see *Table 4*) and mix gently but thoroughly.

(iv)   Pour the starch into the gel mould according to the manufacturer's instructions and allow it to cool for $2-3$ h undisturbed at room temperature. The Buchler vertical gel electrophoresis apparatus (Buchler Instruments Inc., Fort Lee, New Jersey) is used in this laboratory.

To perform the electrophoretic run:

(i)   Thaw cell extracts and centrifuge for $1-2$ min in the microfuge.

(ii)   Carefully remove the 10-slot comb from the mould according to manufacturer's instructions and add $0.03-0.04$ ml of the extracts to the slots using a fine-bore Pasteur pipette or syringe with a 26 gauge needle. Use care to avoid overflow of extracts among adjacent slots.

(iii)   Cover the entire area of the gel, exposed following removal of the slot-former, with heated liquid petroleum.

(iv)   Add buffer to the electrode chambers and secure the loaded gel mould in place with the retention spring (see manufacturer's bulletin).

(v)   Apply the flannelette wicks, place the unit at 4°C and begin the run using $4-5$ V/cm of gel slab.

To visualise the isozymes:

(i)   First remove the starch gel slab, cut off and discard 2 in from both ends and cut the gel horizontally into three identical parts using the gel-cutting-device wire (see manufacturer's bulletin).

(ii)   Separate the three layers and place in staining boxes.

(iii)   Cover the entire surface of the gel layers with the appropriate noble agar-stain mixture and incubate at 37°C in the dark. Bands will appear in $1-3$ h depending on enzyme type and on the activity in the extract.

The zymograms provided in *Figure 4* illustrate typical mobilities which could be expected for isozymes of commonly used lines from a variety of species. See reference (11) for more information and detail.

4.1.3 *Cytogenetics*

Karyologic techniques have long been used informatively to monitor for interspecies contamination among cell lines. In many instances the chromosomal constitutions are so dramatically different that even cursory microscopic observations are adequate. In others, as for example comparisons among cell lines from closely related primates, careful evaluation of banded preparations (10) is required (see Section 4.3.3).

The standard method described below involves the swelling of metaphase-arrested cells by brief exposure to hypotonic saline. The cells are then fixed, applied to slides to optimise spreading, stained and mounted for microscopic observation.

A recommended step-wise procedure is as follows.

(i)   To a culture (T-75) in the exponential growth phase add colcemid to give a final concentration of $0.1-0.4$ µg/ml.

(ii)   Incubate for $1-6$ h, selecting the length of this period roughly by the cycling time of the cell population under study. Diploid human cells with relatively long

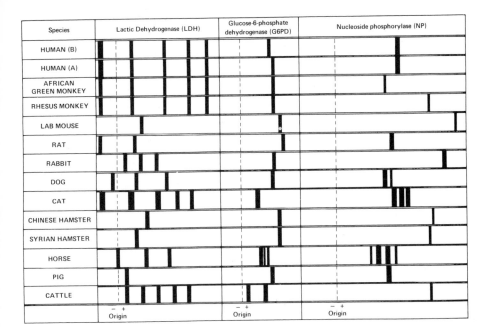

Figure 4. Zymograms for isozymes of lactic dehydrogenase (LDH), glucose-6-phosphate dehydrogenase (G6PD) and nucleoside phosphorylase (NP) showing typical banding patterns of cell lines from some of the more commonly used species.

Table 5. Stock Solutions for Routine Karyology.

Colcemid stock - 1 mg/50 ml DGDW and store frozen in small aliquots.
KCl - 0.075 M in DGDW
Fixative - 3 parts anhydrous methyl alcohol
        1 part glacial acetic acid
        (combine just before use)
Acetic orcein stain - 2 g natural orcein dissolved in 100 ml of 45% acetic acid.
Giemsa stain - 10% in 0.01 M phosphate buffer at pH 7 (solution available commercially).

doubling times would generally require longer incubation than would a rapidly-proliferating Chinese hamster ovary cell line.

(iii)   Gently decant the supernatant and treat the adherent layer with trypsin as for a standard subcultivation. Omit this step if working with a suspension culture. Place the suspension in a 15 ml centrifuge tube. Add serum to a final concentration of approximately 10% of the cell suspension to neutralise further trypsin action on cells.

(iv)   Collect the suspended cells by centrifugation, discard the supernatant and resuspend the pellet in the hypotonic KCl solution at 37°C.

(v)   After 10 − 15 min incubation at 37°C, sediment the cells by centrifugation at 100 g for 10 min and decant the supernatant. Resuspend the cells in a small amount of the hypotonic KCl (~ 1 − 2 times volume of cell pellet).

(vi)   Slowly add 5 ml of freshly-made fixative while agitating the tube manually by pipetting, snapping or using a vortex mixer.

(vii)   After 20 min or more, repeat the centrifugation step, decant the supernatant, add 5 ml of fresh fixative, mix by agitation and let stand at room temperature for $10-15$ min.

(viii)   Repeat step (vii), centrifuge at 100 *g* for 2 min, remove the supernatant and resuspend the cells in a small amount of fresh fixative ($\sim 10-15$ times the volume of cell pellet).

(ix)   To prepare slides, add the suspension dropwise onto clean, cold (4°C) slides held at about a 45° angle, blow gently to spread the cell suspension over the slide and allow to dry completely in air at room temperature.

(x)   Examine the preparation for general arrangement using phase contrast optics. Metaphases should be spread evenly over the slide surface without overlapping. The densities can be adjusted either by altering the number of drops of suspension applied or by changing the concentration of cells in the suspension.

(xi)   Stain with either acetic acid orcein for $3-5$ min or Giemsa for $10-15$ min, rinse in tap water and air dry. Mount in Permount, if so desired.

The frequency of introduction of artifacts through this method will vary depending upon the cell line and the degree of one's experience. Rupturing of cells will occur, for example, and apparent losses or gains in chromosomes will result. However, by counting the chromosomes in $50-100$ well-spread metaphases and recording the modal number the cytogeneticist can obtain a reliable estimate of the true figure for a specific line. Representative data on hundreds of different lines are available in The ATCC Catalogue of Strains II (8).

The karyotype is constructed by cutting chromosomes from a photomicrograph and arranging them according to arm length, position of centromere, presence of secondary constrictions and so forth. Consult the *Atlas of Mammalian Chromosomes* (12) for examples of conventionally stained preparations from over 550 species. For a more critical karyotypic analysis, chromosome banding techniques are required, see Section 4.3.3.

4.2 Tests for Microbial Contamination

The tests included here are suitable for detection of most microorganisms that would be expected to survive as contaminants in cell lines or culture fluids. Techniques for detecting protozoan contamination are not presented as these organisms are rarely found in continuous lines and descriptions of the methodology are available elsewhere (13).

Commercial dry powders are entirely satisfactory for preparation of test media for bacteria, fungi and mycoplasma provided that positive controls are included at least with the initial trials on each lot obtained.

4.2.1 Detection of Bacteria and Fungi

To examine cell cultures or suspect media for bacterial or fungal contaminants proceed as follows.

(i)   Using an inverted microscope, equipped with phase contrast optics if possible, examine cell culture vessels individually. Scrutiny should be especially rigorous in cases in which large-scale production is involved. Check each culture first using low power.

(ii)   After moving the cultures to a suitable isolated area, remove aliquots of fluid from cultures that are suspect and retain these for further examination. Alternatively, autoclave and discard all such cultures.

(iii)   Prepare wet mounts using drops of the test fluids and observe under high power.

(iv)   Prepare smears, heat-fix and stain by any conventional method (e.g., Wright's stain), and examine under oil immersion.

(v)   Consult references 14 and 15 for photomicrographs of representative contaminants and further details.

Microscopic examination is only sufficient for detection of gross contaminations and even some of these cannot be readily detected by simple observations. Therfore, an extensive series of culture tests is also required to provide reasonable assurance that a cell line stock or medium is free of fungi and bacteria. To perform these on stocks of frozen cells:

(i)   Pool and mix the contents of about 5% of the ampoules from each freeze lot prepared using a syringe with a 20 gauge needle. It is generally recommended that antibiotics be omitted from media used to cultivate and preserve stock cell populations. If antibiotics are used, the pooled suspension should be centrifuged at 2000 g for 20 min and the pellet should be resuspended in antibiotic-free medium. A series of three such washes with antibiotic-free medium prior to testing will reduce the concentration of antibiotics which would obscure contamination in some cases.

(ii)   From each pool, inoculate each of the test media listed in *Table 6* with a minimum of 0.3 ml of the test cell suspension and incubate under the conditions indicated. Include positive and negative controls comprising a suitable range of bacteria and fungi which might be anticipated. A recommended grouping consists of

Table 6. Suggested Regimen for Detecting Bacterial or Fungal Contamination in Cell Cultures.

Test medium	Temperature (°C)	Aerobic state	Observation time (days)
Blood agar with fresh defibrinated	37	aerobic	14
rabbit blood (5%)	37	anaerobic	
Thioglycollate broth	37	aerobic	14
	26		
Trypticase soy broth	37	aerobic	14
	26		
Brain heart infusion broth	37	aerobic	14
	26		
Sabouraud broth	37	aerobic	21
	26		
YM broth	37	aerobic	21
	26		
Nutrient broth with 2% yeast	37	aerobic	21
extract	26		

For further detail see reference (15).

Pseudomonas aeruginosa, Micrococcus salivarius, Escherichia coli, Bacteroides distasonis, Penicillium notatum, Aspergillus niger and *Candida albicans.*

(iii) Observe as suggested for 14−21 days before concluding that the test is negative. Contamination is indicated if colonies appear on solid media or if any of the liquid media become turbid.

4.2.2 Mycoplasma Detection

Contamination of cell cultures by mycoplasma can be a much more insidious problem than that created by growth of bacteria or fungi. While the presence of some mycoplasma species may be apparent due to the degenerative effects induced, others metabolise and proliferate actively in the culture without producing any overt morphological change in the contaminated cell line. Thus cell culture studies relating to metabolism, surface receptors, virus-host interactions and so forth are certainly suspect, if not negated in

Table 7. Recipes for Media Used in Culture Test for Mycoplasma.

A. *Stock supplements*

Ingredient	Concentration (g/l)	Ingredient	Concentration (g/l)
Dextrose	50	D-calcium pentothenate	0.024
L-arginine.HCl	10	Pyridoxal.HCl	0.020
Thymic DNA	0.02	Folic acid	0.010
Choline chloride	0.922	Cyanocobalamin	0.003
i-Inositol	0.110	D-biotin	0.002
Niacinamide	0.024	Thiamine.HCl	0.010

Dissolve in DGDW, make to 1 litre, sterilise by filtration, adjust pH to 7.2−7.4 if necessary and store in 100 ml aliquots at −70°C.

B. *Mycoplasma broth*

Ingredient		Instruction
Broth base (Becton Dickinson)	14.7 g	In 600 ml DGDW and autoclave
Phenol red	0.02 g	
Selected horse serum	200 ml	Sterile mix with broth base at room temperature.
Yeast extract, 25%	100 ml	Store at −20°C in 100 ml aliquots at 25%. Final concentration 2.5%.
Stock A	100 ml	

Adjust pH to 7.2−7.4 if necessary, aliquot to sterile test tubes (10 ml/tube), store at 4°C and cap for use within 4 weeks.

C. *Mycoplasma agar medium*

Ingredient		Instruction
Agar base (Becton Dickinson)	23.7 g	In 600 ml DGDW and autoclave. Cool to 50°C in water bath.
Selected horse serum	200 ml	Warm to 37°C and sterile mix with agar base.
Yeast extract (25%)	100 ml	
Stock A	100 ml	

Dispense 10 ml aliquots quickly to 60 mm plates and store at 4°C in a closed container for use within 4 weeks.

Figure 5. Microphotograph showing colonies of mycoplasma (arrows) in an agar plate. The smaller material between colonies is from the inoculum of cultured cells and debris.

interpretation entirely, when conducted with cell lines harboring mycoplasma.

Nine general methods are available for detection of mycoplasma contamination (5,6). Both the direct culture test and the indirect test employing a bisbenzimidazole fluorochrome stain (Hoechst 33258) for DNA are used routinely at the ATCC to check incoming cell lines and all cell distribution stocks for mycoplasma (16).

The serum, yeast extract and other ingredients used for mycoplasma isolation and propagation should be pre-tested for absence of toxicity and for growth-promoting properties before use (5). Positive controls consisting of *M. arginini, M. orale* and *A. laidlawii* are recommended since they are among the most prominent species isolated from cultured cells (6,16).

(i)   To test a culture, be sure that the growth medium is free of antibiotics, especially gentamycin, tylocine or other anti-mycoplasmal inhibitors.

(ii)   Remove all but about 3 ml of fluid from a confluent or dense culture which has not been fed for at least 3 days, and scrape off parts of the monolayer with a rubber policeman. Suspension cultures near saturation may be sampled directly.

(iii)   Inoculate 1.0 ml of the cell suspension to a tube of mycoplasma broth and 0.1 ml onto a plate of mycoplasma agar.

(iv)   Incubate the broth aerobically at 37°C and the agar at 37°C in humidified 95% nitrogen/5% CO_2. A change in the pH of the broth or development of turbidity warns of mycoplasma contamination.

(v)   At weekly intervals for the following 14 days transfer 0.1 ml of the broth to a fresh plate of mycoplasma agar, and incubate at 37°C anaerobically.

(vi)   Using an inverted microscope examine all agar plates at 100 and 300× weekly for a minimum of 3 weeks. A photomicrograph of mycoplasma colonies on agar is provided as *Figure 5*.

Since many strains of mycoplasma, especially *M. hyorhinis*, are difficult or impossible to cultivate in artificial media, an indirect test method should also be included.

To perform the indirect test:

(i)   Inoculate Petri plates (60 mm) containing 10.5 × 22 mm glass coverslips (No. 1)

Table 8. Solutions for the Indirect Fluorochrome Test for Mycoplasma.

Stock Fluorochrome (1000 ×)

Hoechst 33258 stain	5 mg
Thimersol (merthiolate)	10 mg

Mix thoroughly for 30 min at room temperature in 100 ml of HBSS without sodium bicarbonate or phenol red.
Store at 4°C in an amber bottle wrapped completely in aluminium foil. The stain is light and heat sensitive.

Working fluorochrome stock (1−5 ×)

Make up by adding 0.1−0.5 ml of 1000 x stock to 100 ml HBSS without bicarbonate or phenol red and mix again as for 1000 × stock.

Mounting medium

Citric acid (0.1 M)	22.2 ml
Disodium phosphate (0.2 M)	27.8 ml
Glycerol	50 ml

Store at 4°C and check pH just before use. pH 5.5 is optimal for fluorescence.

Fixative

Glacial acetic acid	1 part
Anhydrous methanol	3 parts

---

with 10^5 indicator cells in 4 ml of Eagle's minimum essential medium supplemented with 10% foetal bovine serum. The Vero or 3T6 lines are commonly used. These cells help to amplify low level infections which may otherwise be undetectable.

(ii) After incubating these indicator cultures for 16−24 h at 37°C in an air (95%)/CO_2 (5%) atmosphere, add 0.5 ml of the test cell suspension obtained as outlined for the direct test.

(iii) Return the plates to the incubator for an additional 6 days.

(iv) Remove the culture plates from the incubator, aspirate the fluid and immediately add fixative. It is important not to let the culture dry before fixation as this may introduce artifacts.

(v) After 5 min replace the 'spent' fixative with a fresh volume of the same solution such that the specimen-coverslip is well immersed.

(vi) Aspirate, allow the specimen to air dry and prepare a fresh working stain solution from the 1000 x stain stock.

(vii) Add 5 ml of the staining solution to each dish. Cover and let stand for 30 min at room temperature.

(viii) Aspirate and wash the culture three times with 3−5 ml of *distilled water*, removing the final wash completely to allow the coverslip to dry.

(ix) Remove the coverslip and place *with cells up* on a drop of mounting fluid on a 1 × 3 microscope slide.

(x) Add another drop of mounting fluid on top of the specimen-coverslip and place a larger clean coverslip above this, using a standard mounting technique to avoid trapping bubbles.

(xi) Examine each specimen at 500× using a fluorescence microscope fitted with number 50 barrier (LP440 nm) and BG12 exciter (330/380 nm) filters.

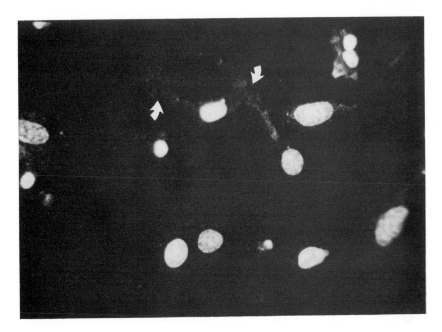

Figure 6. Microphotograph showing mycoplasmal DNA (arrows) demonstrated by Hoechst staining. The larger fluorescing bodies are nuclei of the substrate cell line Vero. (Photograph courtesy of M. L. Macy.)

Note: Confluent cells do not spread sufficiently to stain adequately with Hoechst 33258. Ensure that cultures are subconfluent at the time of staining.

Positive and negative controls should be included with each test series. For this indirect test the Vero or 3T6 cells inoculated with medium only serve as negative controls. *M. hyorhinis, M. arginini, M. orale* and *A. laidlawii* are suitable controls with the order of preference as listed. The nucleic acid of the organisms is visible as particulate or fibrillar matter over the cytoplasm with the cultured cell nuclei more prominent (*Figure 6*). With heavy infections, intercellular spaces also show staining.

Preparation of positive controls, i.e., deliberately infecting cultures, creates a potential hazard to other stocks, so infected cultures should be prepared at a time when, or in a place where, other cultures will not be at risk. Fix and store positive controls at −4°C over desiccant until required.

In all quality control work with mycoplasma or unknown cultures it is prudent to work in a vertical laminar flow hood, preferably isolated from other standard cell culture activity. One should be aware continually of the danger of contaminating clean cultures by aerosols from mycoplasma-containing cultures manipulated in the same area or manually during processing of multiple culture flasks. Appropriate disinfection between uses of a hood or work area is strongly recommended.

Further detail is available in references 5, 6 and 16.

4.2.3 *Testing for the Presence of Viruses*

Of the various tests applied for detection of adventitious agents associated with cultured cells, those for endogenous and contaminant viruses are the most problematical. Development of an overt and characteristic cytopathogenic effect (CPE) will certainly provide an early indication of viral contamination. However, the absence of CPE definitely does not indicate that the culture is virus-free. In fact, persistent or latent infections may exist in cell lines and remain undetected until the appropriate immunological, cytological, ultrastructural and/or biochemical tests are applied. Additional host systems or manipulations, for example, treatment with halogenated nucleosides, may be required for virus activation and isolation.

It should be emphasised at the outset that the protocols presented below represent an expedient compromise established to monitor for readily detectable viruses associated with cell lines. Egg inoculations plus selected co-cultivations and haemadsorption tests are included in addition to routine examinations for CPE using phase-contrast microscopy. Similar general tests are recommended by government agencies in cases where cell lines are to be used for biological production work.

To examine a culture for morphological evidence of viral contamination:

(i)   Hold each flask or bottle so that light is transmitted through the monolayer and look for plaques, foci or areas that lack uniformity. Cultures from frozen stocks should be set up from the pooled contents of about 5% of the ampoules from each lot.

(ii)   Using an inverted microscope, equipped with phase-contrast optics wherever possible, examine cell culture vessels individually, paying special attention to any uneven areas in gross morphology observed previously.

(iii) Prepare coverslip cultures if higher power or additional study is required. The coverslips with monolayers can be fixed and stained by standard histological procedure, and the morphology can be compared with that of appropriate controls.

The presence of certain viruses which do not produce a CPE in cultured cells can be demonstrated by the haemadsorption test. Infected cells in monolayer adsorb indicator red blood cells after brief exposures in the cold. To perform the test:

(i)   Establish test cultures in T-25 flasks using an inoculation density such that the monolayers become confluent in 48 – 72 h. Use pooled cells from about 5% of the ampoules from any given freeze lot to be tested.

(ii)   Remove the fluid from confluent cultures and rinse the test monolayers with 5 ml HBSS minus divalent cations.

(iii) Add 0.5 ml each of freshly-prepared suspensions of chick, guinea pig and human type O erythrocytes (0.5% v/v, washed three times in succession with saline). Then place the flask with monolayer down at 4°C for 20 min.

(iv)   Observe macroscopically and microscopically under low power for clumping and adsorption of red blood cells to the monolayer.

(v)   Repeat on all test cultures not exhibiting haemadsorption before recording a negative result.

A suitable positive control can be established by infecting a flask of RhMK with 0.2 ml of undiluted influenza virus 48 – 72 h before testing.

The inoculation of test cells or cell homogenates into embryonated chicken eggs provides an additional sensitive test for the presence of viruses. To prepare the eggs for this procedure:

(i)   Drill a small hole in the air sac (blunt end) of an 8 − 9 day embryonated egg using an electric drill and a 1/16 inch burr-type bit. In this and subsequent operations, work with sterile instruments. Swab areas of the shell to be drilled with 70% ethanol before and after each manipulation. The drill bits may be placed in 70% ethanol before use.

(ii)   Using an egg candling lamp, locate the area of obvious blood vessel development, and at a central point carefully drill through the shell leaving the shell membrane intact.

(iii)   Place 2 or 3 drops of Hank's balanced salt solution (HBSS) on the side hole and carefully pick through the shell membrane with a 26 gauge syringe needle. The saline will seep in and over the chrorioallantoic membrane (CAM) to facilitate its separation from the shell membrane.

(iv)   Apply gentle suction to the hole in the air sac using a short piece of rubber tubing with one end to the mouth and the other pressed to the blunt end of the egg. Use the candling lamp to monitor formation of the artificial air sac over the CAM.

(v)   Seal both holes with squares of adhesive or laboratory tape and incubate the eggs horizontally at 37°C. Standard cell culture incubators and walk-in rooms are entirely adequate for egg incubations. High humidity or air/CO_2 boxes are not satisfactory.

The embryonated eggs with artificial air sacs are ready to receive the test cells or homogenates in 24 − 48 h. It is important to examine the eggs prior to injection and discard any which show signs of death or degeneration. The presence of intact blood vessels and embryo movements generally indicate that the preparation is healthy. In this event, proceed as follows:

(i)   Obtain suspensions of test cells in the appropriate growth medium and adjust the concentration such that 0.2 ml contains $0.5 − 1 \times 10^7$ cells. If the cells are or may be tumourigenic, freeze in a dry-ice alcohol bath and thaw quickly in a water bath at 37°C. Repeat twice.

(ii)   Remove the seal from the side holes in the embryonated eggs and inject 0.2 ml of the cell suspension or homogenate onto the CAM of each of 5 − 10 eggs.

(iii)   Using the candling lamp, examine the embryos 1 day after adding the cell suspension; discard any embryos that have died. Repeat the examination periodically for 8 − 9 days.

(iv)   If embryos appear to be viable at the end of the incubation period, open the eggs over the artificial air sac and examine the CAM carefully for oedema, foci or pox. Check the emryo itself for any gross abnormalities such as body contortions or stunting.

(v)   In cases in which viral contamination is indicated, repeat all steps both with a second aliquot of the suspect cells and with fresh fluid samples from eggs in which the embryos have died or appear abnormal.

Positive controls may be established by inoculating eggs with influenza virus, Newcastle disease virus and/or Rous sarcoma virus. Negative controls receive only an injection of standard cell culture medium. Consult reference (17) for diagrams and more detail on egg inoculations.

The co-cultivation of test and presumptive-substrate cell lines provides another suitable method for detection of viruses, and this technique is especially useful for suspension cell cultures. The substrate or target line of choice will depend upon the species from which the test cell line originated. For example, for human cell lines, one could co-cultivate with WI-38, MRC-5 or primary human embryonic kidney cells (HEK). A cell line from a second species of choice in this example could be the African green monkey.

After selecting appropriate substrate lines, initiate and maintain the co-cultivation as follows.

(i)   Inoculate a T-75 flask with 10^6 cells from each line in a total of 8 ml of an appropriate growth medium. In some cases the inocula may have to be adjusted in an attempt to maintain both cell populations during the co-cultivation period. For example, if a very rapidly proliferating line is co-cultivated with a test line that multiplies slowly, the initial ratio of the former to the latter could be adjusted to 1:10. Similarly, the population that multiplies slowly might have to be re-introduced to the co-cultivation flasks if it were being overgrown by the more rapidly dividing cells.

(ii)   Change the culture fluid twice weekly and subcultivate the populations as usual soon after they reach confluence. If the test line grows in suspension, that population will have to be recovered by centrifugation and subcultivated by dilution as usual.

(iii)   Examine periodically for CPE and haemadsorption over a 2 − 3 week period at minimum, using procedures described above.

Viral isolates may be identified through standard neutralisation (haemadsorption inhibition, plaque inhibition, haemagglutination inhibition) or complement fixation tests.

It should be emphasised that, in spite of these screens, latent viruses and viruses which do not produce overt CPE or haemadsorption will escape detection. Consult references 17 and 18 for other specialised tests for contaminant viruses and for more detail.

4.3 Testing for Intraspecies Cross-contamination

With the dramatic increase in numbers of cell lines being developed, the risk of intraspecies cross-contamination rises proportionately. The problem is especially acute in laboratories where work is in progress with the many different cell lines of human and murine origin (hybridomas) available today. Tests for polymorphic isoenzymes, surface marker antigens and unique karyology are all important tools to detect cellular cross-contamination within a given species.

4.3.1 *Isoenzymology*

A method for the verification of cell line species by determining mobilities of G6PD, LDH and NP isozymes is outlined in Section 4.1.2. By using similar technology one can also screen for intraspecies cross-contamination.

In cell lines from various individuals of the same species there are often different co-dominant alleles for a given enzyme locus, the products of which are polymorphic

Table 9. Procedures[a] for ADA, ESD, PEP-D, PGM's and PGD.

Enzyme	Chamber buffer	Starch buffer	Staining solutions
ADA	TC	TC (0.1 ×)	20 ml 0.1 M Na phosphate, pH 7.0. 30 ml H_2O 50 mg adenosine 3 mg PMS 10 mg MTT 50 μl nucleoside phosphorylase (1 mg/ml) 50 μl xanthine oxidase (10 mg/ml)
ESD	TC	TC (0.1 ×)	50 ml 0.05 M Na acetate pH 5.2 10 mg methyl umbelliferyl acetate dissolved in 2 ml acetone
PEP-D	TEB	TEB (0.1 ×)	40 ml H_2O 10 ml 0.1 M $NaH_2PO_4.H_2O$ pH 7.4 20 mg L-phenylalanyl-L-proline 10 mg horseradish peroxidase (type 2) 50 μl amino acid oxidase (1 mg/ml) 2 mg o-dianisidine 0.2 ml 0.1 M $MnCl_2$ (add just prior to mixing with overlay)
PGM $_1$ and $_3$	TC	TC (0.1 ×)	35 ml H_2O 15 ml 0.5 M Tris pH 8.0 200 mg glucose-1-phosphate with 1% glucose-1,6-diphosphate 25 mg $MgCl_2$ 3 mg PMS 10 mg MTT 100 μl glucose-6-phosphate dehydrogenase (100 U/ml)
PGD	TEB	TEB to (0.1 ×)	35 ml H_2O 15 ml 0.5 M Tris pH 7.0 200 mg 6-phosphogluconate, Na 20 mg NADP 3 mg PMS 10 mg MTT

[a]All gels are run for 16−18 h at 4°C at 4−5 V/cm (160−180 V). NADP is added (5 ml at 0.005 M) to the cathode buffer as well as to the staining solution in each case where an NADP-dehydrogenase is being assayed. See *Table 4* and related text for further instructions.
[b]For all enzymes except ESD, these solutions are mixed just before use with an equal volume of 2% noble agar in water which has been liquified and cooled to 45−50°C. For ESD, a filter paper is placed over the gel and the solution is poured onto the paper. In this latter case, the allozyme locations are visualised by removing the filter paper after incubation and observing the gel under long wave ultraviolet illumination. Bands appear within 10 min to 3 h depending upon cell and allozyme type, and density of cells used in the homogenate.

and electrophoretically resolvable. In most cases the phenotype for these allelic isozymes (allozymes) is extremely stable. Consequently when allozyme phenotypes are determined over a suitable spectrum of loci they can be used effectively to provide an allozyme genetic signature for each specific line under study (19−24).

With human cell lines, the definition of allozyme phenotypes for seven or more enzymes, for example, adenosine deaminase (ADA); G6PD; esterase D (ESD); peptidase-D (PEP-D); phosphoglucomutases 1 and 3 (PGM$_1$, PGM$_3$) and 6-phosphogluconate dehydrogenase (PGD) has been shown to be sufficient to provide identification with a high degree of confidence (20). Methodology as described in Section 4.1.2 can be

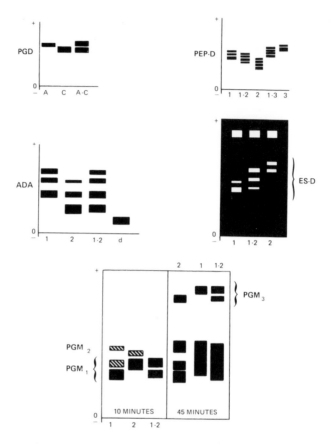

Figure 7. Zymograms showing representative phenotypes for six polymorphic enzymes of human cell lines. PGD, 6-phosphogluconate dehydrogenase; PEP-D, peptidase-D; ADA, adenosine deaminase; ES-D, esterase D; $PGM_{1 3}$, phosphoglucomutases 1 and 3. The origin is indicated in each case (0) with some of the more common phenotypes (abscissae). Multiple allozyme bands may be apparent in some cases or they may migrate so closely that they appear as a single, more dense band. The PBM zymograms are of the same gel at two different times since PGM_3 develops later. PGM_2 develops at an intermediate time and complicates interpretation of PGM_1 bands.

used. The appropriate staining solutions are listed in *Table 9*.

Example zymograms for six of these allozymes for human cell lines are given as *Figure 7*. For ADA, only the red cell forms are shown for simplicity. In human cells from some tissues additional forms of ADA have been observed. These may represent secondary ADA isozymes produced by post-translational modification of the red cell ADA isozymes. Furthermore, expression of ADA isozymes is affected by culture conditions (23).

PGM_3 allozymes are the most rapidly migrating forms of that group and are generally much lower in activity and therefore later staining. PGM_1 allozymes are slowest in migration, stain rapidly and can be visualised best at an early stage before PGM_2 and PGM_3 are more fully developed.

The frequency of occurrence of the genetic signature for any particular cell line can

Table 10. Phenotype Frequencies of Seven Polymorphic Enzymes for White and Black Human Populations[a].

Enzyme	Phenotype	Phenotype frequency	
		Whites	Blacks
ADA	1	0.8845	0.9320
	1−2	0.1138	0.0668
	2	0.0017	0.0012
G6PD	A	0	0.4400
	B[b]	1	0.5600
ESD	1	0.8131	0.8039
	1−2	0.1776	0.1961
	2	0.0092	0.0096
PEP-D	1	0.9780	0.9110
	1−2	0.0220	0.0479
	1−3	0.0000	0.0411
	2	0.0001	0.0006
	3	0.0000	0.0005
PGM$_1$	1	0.5849	0.6019
	1−2	0.3565	0.3204
	2	0.0558	0.0485
PGM$_3$	1	0.5403	0.1362
	1−2	0.3997	0.4085
	2	0.0600	0.4553
PGD	A	0.9536	0.9264
	AC	0.0455	0.0727
	C	0.0009	0.0009

[a]Excerpted from reference (22).
[b]Male data only; G6PD is X-linked.

be estimated from the allozyme phenotype observed and data on phenotype frequencies available for normal human populations (*Table 10*). Thus, for example, a cell line with genetic signature ADA-1, G6PD-B, ESD-1, PEP-D 1, PGM$_1$-2, PGM$_3$-1, and PGD-A derived from a white male would have a phenotype frequency product of 0.8845 × 1.0 × 0.8131 × 0.9780 × 0.0558 × 0.5403 × 0.9536 = 0.0202. This figure is useful in that it reflects the probability of encountering a line with that particular signature and is therefore informative when questions of cross-contamination arise (20,22).

The ability to discriminate identities among cell lines using this method can be increased of course, by determining the profiles of additional allozymes. Some of the more useful for this purpose include acid phosphatase (ACP1), glyoxalase 1 (GLO1), malate dehydrogenase (ME2), alpha fucosidase (FUCA) and adenylate kinase (AK1) (22,24).

Polymorphic isozymes have also been described for murine cell lines and hybrids. Allozymes of potential use in monitoring for intraspecies cross-contamination in this case include the esterases (ES-1 and ES-2); dipeptidase 1 (Dip-1); glutamic-oxaloacetic transaminase 2 (GOT-2); isocitric dehydrogenase 1 (NADP-dependent form); NADP-malic enzyme (Mod-1, supernatant form); glucose-phosphate isomerase 1 (GPI-1) and phosphoglucomutases 1 and 2 (PGM$_1$ and $_2$). While methodology for their electrophoretic separation is available (25) these allozymes have not yet been fully exploited in the development of signatures for specific murine cell lines.

4.3.2 *Tests for Blood Group and Histocompatibility Antigens*

The blood group and human leukocyte antigens (HLA) on the plasma membrane of human cells in culture provide additional, highly useful markers for identification. Lack of expression or partial expression of these has been documented in a number of cases.

Blood group antigens are present on normal human epithelia in primary culture (26) and on some continuous epithelial lines. The standard test for these can be applied as follows.

(i)   Obtain a cell suspension by dissociation or harvest and wash twice in PBS by centrifugation with subsequent resuspension as usual.

(ii)   Perform a cell count and adjust the final cell concentration to $1-2 \times 10^6$/ml.

(iii)   Place a drop of the suspension on each of four separate locations on a large microscope slide and add a drop of anti-human A, B or AB typing antiserum to each. Add a drop of PBS to the fourth drop of cell suspension to provide a negative control.

(iv)   Immediately mix each pool separately with glass rods and observe under low power for agglutination. The negative control may also show some cell clumping but this should be minor when compared with the positive test suspension-antiserum mix.

Experience in this laboratory indicates that in many cases the donor's blood group type is expressed even on lines from malignant tumors (8,9). Not infrequently, however, cells of other human lines from A,B or AB donors will not react, thus giving a false negative, type O reading. The hypothesis that this could be due to removal of the antigen during dissociation has been tested by repeating the assay on cells maintained in suspension culture for $2-18$ h. In spite of the fact that this should allow for replacement synthesis of the surface antigen, inappropriately negative lines remained negative.

This problem with blood group antigen detection on cultured epithelia should be recognised. The simple test, coupled with others for intraspecies contamination among lines, is valuable nevertheless in initial screening. Example results are recorded for a variety of lines in references (8) and (9).

The major histocompatibility system in humans consists of HLA antigens present on the plasma membrane of most nucleated cells. These numerous antigens, which are coded by co-dominant genes (77 alleles) of five closely linked loci on chromosome 6, provide one of the most polymorphic human group systems known. The antigens are detected routinely by a two-stage, complement-dependent cytotoxicity test, and dye-exclusion is used to estimate loss of viability (Section 3.3.1).

A simplified and standard outline of the procedure follows.

(i)   Load wells of plastic histocompatibility typing plates at 1 μl/well with the desired panel of antisera using a Hamilton dispenser. Antisera, which are available commercially, are generally obtained from individuals immunised to HLA antigens by pregnancy or blood transfusions. Typing plates loaded with a spectrum of HLA antisera can also be purchased for routine clinical work. Negative controls should be included with each cell line and run.

(ii)   Obtain a cell suspension, wash as for the blood group antigen assay using medium RPMI 1640 without serum and adjust the cell concentration such that 1 μl with

10^5 cells can be added to each well using a single place Hamilton dispenser.

(iii)   Mix by placing the typing plate against a Yankee pipette shaker and incubate at 20°C for 30 min.

(iv)   Add 5 μl of rabbit complement to each well. Incubate at 20°C for 60 min.

(v)   Dispense 3 μl of 5% aqueous eosin to each well. Wait 2 min for dead cells to stain.

(iv)   Fill wells with buffered formalin (pH 7.2) and add a cover glass (50 × 75 mm) to flatten the droplets.

(vii)   Observe and record the incidence of staining using an inverted microscope. The degree of staining is usually by approximation rather than actual cell count.

While this procedure can be applied successfully for typing some cell lines, it should be emphasised that modifications will be required in many individual cases. The major variables are non-specific antibodies present in the rabbit complement or HLA antisera.

Rabbit serum is the most satisfactory source of complement for this reaction due to the presence of natural antibodies to human cells. The interaction of these with other cell surface antigens enhances the complement-dependent cytotoxic effect of the anti-HLA antibody-antigen union. The titres and specificities of these natural antibodies differ even among pooled rabbit sera. The problem may be overcome by varying the incubation times, by trying different sources and dilutions of complement, by diluting the rabbit serum with human serum or by absorption of the rabbit serum with cultured cells (27). Each cell line may have to be examined separately since the end result depends upon the multiple interactions between antibodies present with the spectrum of antigens on the surface of each cell type.

The presence of given HLA allo-antigens on a particular cell line can be confirmed by absorption-inhibiting typing. In this case the ability of the cells to absorb HLA allo-antibodies from antisera of known specificity is determined by quantitating the loss of cytotoxic effect after absorption. To accomplish this, the pre- and post-absorption antisera are titrated in parallel against a panel of lymphocytic lines of known HLA profile (27).

4.3.3 *Giemsa Banding*

A powerful method for cell identification which is close to absolute in some cases, involves karyotype analysis after treatment with trypsin and the Giemsa stain (Giemsa or G-banding). The banding patterns made apparent by this technique are characteristic for each chromosome pair and permit recognition by an experienced cytogeneticist even of comparatively minor inversions, deletions or translocations. Many lines retain multiple marker chromosomes, readily recognisable by this method, which serve to identify the cells specifically and positively.

Many modifications of the original technique (28) are available. The following can be applied to metaphase spreads obtained as indicated in Section 4.1.3 [step (x)].

(i)   Use air-dried slides within 2 − 7 days.

(ii)   Incubate in 0.025 M phosphate buffer (PB) (3.4 g KH_2PO_4/l adjusted to pH 6.8) at 60°C for 10 min. The PB is also used in subsequent steps.

(iii)   Prepare a staining solution *just before use* consisting of 6.5 ml PB, 0.55 ml of a trypsin solution (1% Difco 1:250 distilled water), 2.5 ml of 100% methanol and 0.22 ml of stock Giemsa solution (commercial). The 1% trypsin stock may be made in bulk and stored in aliquots at $-70°C$.

(iv)   Flood each slide with about 1 ml of staining solution and leave for 15 min.

(v)   Rinse briefly with distilled water and air dry completely.

Completely dried slides are used to examine under brightfield, oil-emersion plana-pochromat objectives without coverslip. Oil can be placed directly on the slide. However, care must be taken not to scratch the cell surface. Oil must be removed as completely as possible immediately after the use of slides. Generally a few changes in xylene should be satisfactory. A typical banded preparation is shown as *Figure 8*.

See references (29) and (30) for further detail and additional example preparations.

4.3.4 *Other Methods*

New technique are emerging which appear to offer high potential for cell identification especially at the more critical intraspecies level. The methodology will not be outlined here but is discussed further in references (31) and (32).

One of these methods involves use of high-resolution two-dimensional electrophoresis as reviewed by the Andersons (31). The technique depends on the use of two different

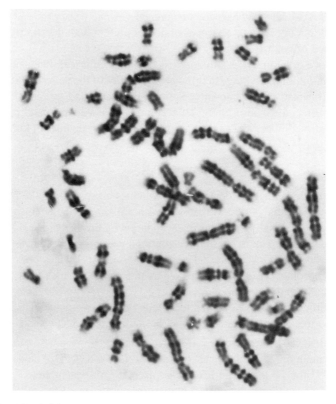

Figure 8. Giemsa banded chromosomes from the cell line ATCC.HTB 16 (U-138 MG) isolated from a human glioblastoma. (Photograph courtesy of T. R. Chen.)

separations, which rely on two unrelated parameters, the isoelectric point and molecular weight. The analytical methods used to achieve these two separations, which are isoelectric focusing in the presence of urea and electrophoresis in the presence of sodium dodecyl sulphate, each have resolutions in practice of over 100 proteins. When combined two-dimensionally, the resolution of the overall system is in theory over 10 000, and in practice well over 2000 for a single analysis. Not all proteins are seen on the standard gels, because some of the basic and acidic proteins are lost on the edges. Non-equilibrium gradient techniques have therefore been developed which give 1000 or 2000 more spots for some cell types.

A second method requires application of recombinant DNA technology and cloned DNA probes to identify and quantitate allelic polymorphisms. To date, over 100 DNA polymorphisms in the human genome have been recognised and chromosomally assigned, and this number is increasing rapidly. These polymorphisms can provide extremely useful markers even if they are not expressed through transcription and translation to yield structural or enzymatically active proteins. A screening for these markers by Southern blot analyses has already been used, albeit on a small scale, to demonstrate utility in cell line identification (32).

Perhaps within the next decade all lines in major use will have been definitively typed by one or both of these techniques.

4.4 Verifying Tissue of Origin

The markers used for verification of the source tissues for cell lines are probably as numerous as are the types of metazoan cells. The means of major utility include analyses of fine structure, immunological tests for cytoskeletal and tissue-specific proteins, and of course any of an extremely broad range of biochemical tests for specific functional traits of tissue cells. Some of the more general methods in these categories will be outlined below.

4.4.1 Fine Structure Analyses for Characteristic Markers

The presence of desmosomes and keratin filaments, both of which can be visualised by electron microscopy, is generally accepted as characteristic of epithelia. The so-called Weibel-Palade bodies are specific for endothelial cells of the umbilical vein and other sources. Cells of the islets of Langerhans can be characterised morphologically by demonstrating the presence of their specific secretion granules.

A basic technique to prepare cultured cells for such fine structural analysis is outlined below. Reagents required are listed in *Table 11*. To prepare a cell monolayer or pellet of cells harvested either from a suspension culture or by dissociation:

(i)   Collect a pellet by centrifugation of not more than $1-3 \times 10^6$ cells in a 15 ml glass or polypropylene test tube. Monolayers can be processed directly on plastic culture vessels.

(ii)   Remove culture medium and wash the cells once with stock buffer at 4°C. All steps through (v) are performed at this temperature.

(iii) Fix for 1 h in 2% buffered glutaraldehyde. If a pellet is being processed dislodge it gently from the test tube sides using an orange stick, thus facilitating diffusion of reagents from all sides of the three-dimensional mass. It is important to retain

Table 11. Reagents[a] used in Processing Cultured Cells for Electron Microscopy.

Stock solutions

1. Sodium cacodylate buffer, 0.1 M, pH 7.2−7.3
    Na cacodylate   42.8 g
    Aqueous phenol red (0.5%)   4 ml
    Dissolve in 1 l DGDW and adjust pH with 1 M HCl
2. Glutaraldehyde: 8% aqueous
    Dilute to 4% with DGDW
3. Osmium tetroxide 4% aqueous
4. Uranyl acetate 0.5% aqueous
5. Reynolds' lead citrate
    Lead nitrate   1.33 g
    Trisodium citrate, dihydrate   1.76 g
    DGDW   30 ml
    Dissolve with stirring for about 30 min. Add 8.0 ml of 1 M NaOH (carbonate free)
    and dilute to 50 ml.

The solution, which is stable for at least 6 months should be passed through a Millipore filter (0.22 μm or less) just before use to avoid small particular precipitates.

6. Alcohol series to 100% absolute and propylene oxide
7. Plastic mixtures
    A. Epon 812   62 ml
        Dodecenyl succinic anhydride (DDSA)   100 ml
    B. Epon 812   100 ml
        Methyl nadic anhydride (MNA)   89 ml

Thes can be made up in advance and stored at 4°C in well-sealed containers for up to 6 months. Accelerator (2,4,6-tri dimethyl-aminomethyl phenol) (DMP 30) is added just before use at 1.5−2.0%.

Working solutions

1. Buffered glutaraldehyde (2%) 1:1 stock buffer/4% glutaraldehyde stock
2. Buffered osmium tetroxide (1%) 1:3 4% osmium/stock buffer
3. Plastic mixture

Thoroughly mix 7 parts of mixture B with 3 parts of mixture A just before use and add the accelerator at 1.5−2.0%.

---

[a]Can be obtained from Polysciences, Inc., Warrington, PA.

a mass of cells throughout as opposed to a dispersed population. These larger conglomerates will pass through the more viscous mixtures to be used in later steps.

(iv)   Decant off the fixative and rinse three times with buffer for at least 10 min per rinse.

(v)   Post-fix with 1% osmium tetroxide for 1 h.

(vi)   Decant the osmium tetroxide fixative and wash three times with double glass-distilled water (DGDW) (5 min each).

(vii)   Leave for 16−20 h in 0.5% aqueous uranyl acetate. The low pH of this solution removes glycogen but gives excellent membrane preservation and staining.

(viii)   Pass through an alcohol series (10-30-70-95-100-100%) leaving the sample for 10 min in each solution. For pellet fragments, three subsequent 10 min infiltrations with 1:1 100% alcohol/100% propylene oxide; 100% propylene oxide and 1:1 propylene oxide/complete plastic mixture are recommended. These latter three steps cannot be used with monolayers on plastic. In that case, infiltrate directly from 100% alcohol for 30 min with a 1:1 solution of the complete plastic mixture and 100% alcohol.

(ix) Decant and begin infiltration with 100% complete plastic mixture (7:3 B/A plus 1.5 − 2.0% DMP-30, see *Table 11*). For pellet fragments this can best be accomplished by placing a fragment no larger than 1 mm in diameter in the upper part of the mixture in a BEEM embedding capsule (Polysciences). The fragment should settle to the bottom (pointed end) of the capsule within a few hours.

For a monolayer the complete plastic mixture is placed in the culture vessel and allowed to infiltrate for 8 − 18 h.

Bubbles can be removed by placing the BEEM capsules or open culture vessels in a vacuum.

(x) Polymerise at 65 − 70°C for 24 h or longer to give a suitable hardness. The B/A mix is rather hard and can be modified depending on the technician's experience and preference (33).

(xi) For infiltrated monolayers, cut out the area(s) to be examined using a fine jeweller's saw, and cement to a dummy plastic rod. An 'instantly' drying form of contact cement is best. The sample can be oriented for cross-sectioning or to collect sections which are tangential to the monolayer surface (e.g., for desmosomes). For pellet fragments, simply snap off the BEEM capsule, trim as necessary with a single edged razor blade and place in the microtome chuck.

(xii) Section using an ultramicrotome (see manufacturer's instructions). Silver or grey sections are suitable for examination at magnifications above 15 000. Gold sections can be used at lower magnifications.

(xiii) Collect the sections on a suitable grid, stain for 20 − 60 sec, in lead citrate, examine and photograph as desired. Consult references (33) and (34) for further information.

4.4.2 *Immunological Testing for Cytoskeletal Proteins*

The cytoplasmic intermediate-filament (IF) proteins which form an essential part of the cytoskeleton have emerged as being highly useful in cell identification. At least five filament subgroups have been described on the basis of morphology, polypeptide composition and unique immunogenicity. The distribution of these has been shown to be tissue-type specific. Thus epithelia from most tissues contain cytokeratin(s), and cells of mesenchymal tissues contain vimentin. Desmin is present in myogenic cells; neurofilament protein in neurons and glial fibrillar acidic protein in glial cells (35).

The filaments may be visualised by electron microscopy, by direct or indirect immunofluorescence staining or one can demonstrate specific cytoskeletal protein by immunohistochemical means as indicated below for keratin. Reagents (*Table 12*) and staining kits for the latter are available from a number of commercial sources.

To test for the presence of keratin treat coverslip cultures or cells collected on a slide by cytospin as follows.

(i) Fix for 20 min in methanol:3% H_2O_2 and wash briefly (~30 sec) in a bath of running tap water.

(ii) Place in wash buffer for about 5 min, remove and drain but carefully avoid drying here and throughout the procedure.

(iii) Cover the monolayer or cytospin 'spot' with buffered normal serum (*Table 12*) and leave at room temperature for 30 min.

Table 12. Reagents for the Immunohistochemical Demonstration of Keratin[a].

1. Fixative	50 ml 3% hydrogen peroxide
	250 ml anhydrous methanol.
	Prepare fresh just before use
2. Tris buffer	0.1 M, pH 7.6 used for 3−6 and 8
3. Normal serum	10% swine serum in Tris buffer
4. Rabbit anti-keratin antiserum	1:20 to 1:100 in Tris buffer
5. Swine anti-rabbit IgG antiserum	1:20 in Tris buffer
6. Horseradish peroxidase-rabbit anti-horseradish peroxidase-soluble complexes (HPAP)	1:100 in Tris buffer
7. Wash buffer	1 part Tris buffer (0.5 M, pH 7.6)
	9 parts 0.9% normal saline
8. Chromogen	6 mg 3,3'-diaminobenzidine tetrahydrochloride
Prepare fresh just before use	0.1 ml 3% hydrogen peroxide
	10 ml Tris buffer

[a]A kit adapted from this method (36) is available from DAKO Co., Santa Barbara, CA.

(iv)   Drain off serum, add the diluted rabbit anti-keratin antiserum and retain at room temperature for 30 min.

(v)   Drain, wash gently several times with wash buffer to remove unbound antiserum, place in a coplin jar containing wash buffer and leave for 15 min.

(vi)   Drain, cover the cells with swine anti-rabbit IgG antiserum and leave for 30 min.

(vii)   Rinse several times with wash buffer, place in a coplin jar with fresh buffer for 15 min, and drain carefully.

(vii)   Cover cells with the buffered HPAP reagent, leave at room temperature for 30 min, rinse several times in wash buffer and place in a coplin jar with fresh buffer for 15 min.

(ix)   Drain, cover with the freshly-prepared chromogen solution and allow to stain for 5 min.

(x)   Wash gently with DGDW, counterstain with hematoxylin, dehydrate and mount in permount.

A positive reaction is indicated by the localised deposit of dark-brown, precipitated reaction product. Negative controls which are run in parallel with the test samples, consist of slides with cell preparations exposed to buffer in step (iv) instead of anti-keratin antiserum. Positive controls consists of carcinoma cell lines such as SCC4 or SCC15. Alternatively, sections of skin serve well. See references (35) and (36) for further detail.

4.4.3 *Demonstration of Tissue-specific Antigens*

An array of reports is available describing the isolation and characterisation of tissue and tumour-specific antigens. The methodologies vary somewhat among laboratories and with the different cells lines or tissues used. Only a few examples will be summarised to provide perspective.

The human lung tumour line 2563 (derived from MAC-21) was used directly by Akeson (37) as the immunogen in rabbits. The resulting antiserum was found by complement fixation, immunofluorescence and saturation binding assays to contain antibodies specific for antigens of the normal kidney tissue. These could be selectively absorbed out by kidney preparations showing the existence of a distinct additional antigen shared by the two normal organs.

Ceriani and associates present data demonstrating the existence of mammary-specific antigens on human normal and tumourigenic breast cells. These workers immunised rabbits with a defatted preparation from the washed cream fraction of human milk. Species-specific antibodies were removed from the resultant antiserum by absorption with human erythrocytes. Indirect immunofluorescence studies demonstrated that antibodies in this antiserum bound specifically to cell lines from normal human breast and breast carcinomas. A sensitive radioimmunoassay was developed which permitted quantitative confirmation of this finding. Seven breast carcinoma lines and primary cultures of mammary epithelia bound at least $10-40$ times as much antibody as did breast fibroblasts or epithelia from other sources. Trypsinisation of MCF-7 or MDA-MB-157 removed $80-85\%$ of the reactivity indicating that the breast-specific antigens reside on the cell surface (38).

Procedures were developed by Chu *et al.* for the isolation, purification and quantitation of prostate antigen (PA) from human seminal fluid or prostatic tissue. Crude extracts were treated with ammonium sulphate, the precipitates were solubilised, dialysed and further purified by column chromatography (DEAE and Sephadex). Fractions in molecular weight regions $26\,000-37\,000$ which contained the PA were concentrated by ultrafiltration and this final preparation was used to immunise rabbits.

Interestingly, the resultant antiserum showed a single reaction with prostate tissue extracts on immunodiffusion and did not react with similar preparations from other tissues. Furthermore, quantitative studies, accomplished after development of a sensitive enzyme immunoassay for PA, indicated that human cell lines from prostatic carcinomas (LNCaP and PC-3) expressed antigen. In contrast, cell lines from other tissues had no detectable PA.

Hybridomas producing monoclonal antibodies which show a high degree of tissue specificity have been isolated and characterised in a number of laboratories. For example, Koprowski *et al.* (40) reported on the isolation of 19 hybridomas, 15 of which produced antibodies which were specific for cell lines from colorectal carcinomas. Metzgar and associates (41) developed five hybridomas producing antibodies to human pancreatic tumour cell lines. Some cross-reactivity was noted with other tumour tissues or cell lines but at least in one case (with DU-PAN-1) this was restricted to a transitional cell carcinoma of the bladder. Hybridomas producing antibodies of utility in distinguishing human B cell, T cell and T cell subsets have also been described (42), and many reagents and kits for this purpose are already available commercially.

Because of the unlimited potential availability of monoclonal reagents one might anticipate that kits for identification of cells from many tissues will be developed for laboratory use. These may ultimately replace heterogeneous antisera currently in use for such purposes.

4.4.4 *Biochemical Testing for Cell-specific Function*

A large number of cell lines and strains can be shown to derive from a particular tissue or tumour by the presence of specific synthetic abilities or metabolic pathways. The methodology for such verification of identity is too extensive to be covered in detail. Only a few examples are cited of strains and lines which express specific function.

Yasumura and associates (43) used transplantable mouse and rat tumours as source material for the establishment of four, clonally derived strains of functional cells. Alternating periods of transplantation in host animals and maintenance in culture were required initially to promote growth during isolation. One strain designated GH_1, which was obtained from the MtT/W5 pituitary tumor releases growth hormone into the culture medium and promotes host growth on re-inoculation to either intact or hypophysectomised animals. A second strain, I-10, was derived from the Leydig cell tumor H10119 maintained in BALB/cJ mice. Progesterone and its derivative 20 alpha-hydroxy-4-pregnen-3-one are the major steroids secreted by these cells in culture. They do not respond to interstitial cell-stimulating hormone but are stimulated by cyclic AMP. The murine cultured cell strains M-3 and Y-1 were isolated from a Cloudman S91 melanoma and an adrenal tumour, respectively. The M-3 cells produce melanotic tumors upon re-inoculation into suitable hosts (CXDBA animals). The Y-1 adrenal cell strain releases steroid hormones and responds to ACTH.

The clonal isolation technique has also been applied successfully for the derivation of functional cell strains from normal tissues. One excellent example is FRTL, a rat thyroid cell strain that secretes thyroglobulin into the culture medium and concentrates iodide 100-fold (44). Medium F-12M supplemented only with $0.1 - 0.5\%$ calf serum, insulin, thyrotropin, transferrin, hydrocortisone, somatostatin and glycl-L-histidyl-L-lysine acetate was used. The low serum conditions were said to be critical during early stages for selection of the functional cell type. Strains of myogenic cells have also been isolated from normal rat and mouse muscle. In these cases the specific activity of creatine phosphokinase can be used as a biochemical marker (45).

Functional lines are also available from a variety of human tumours. MCF7, for example, was isolated from the pleural effusion of an individual with adenocarcinoma of the breast. The line retains several characteristics of mammary epithelium including the abilities to process oestradiol *via* cytoplasmic oestrogen receptors and to form domes in confluent monolayers (46). BeWo is a human trophoblastic cell line isolated from a malignant gestational choriocarcinoma of a foetal placenta. Interestingly, the line has been shown to secrete a spectrum of placental hormones including human chorionic gonadotropin, placental lactogen, plus oestrone, oestradiol, oestriol and progesterone (8). See references (7), (8) and (47) for additional detail on differentiated lines and biochemical characterisations.

4.5 Identifying and Quantitating Immunological Products

Given the current intense interest and activity in hybridoma technology, no chapter on cell line characterisation would be considerd complete without reference to this area of endeavour. The specificity of monoclonal antibody produced by an hybridoma is, of course, the ultimate criterion on which identity must be based. If this cell product is absent and its synthesis cannot be restored, the hybrid cell strain is of little use. However, the potential number of different hybridomas which could be generated theoretically equals the number of epitopes which exist times the number of different fusion lines available. Clearly for a hybridoma banking agency or department this presents a dilemma since no single organisation can realistically identify all monoclonal antibodies even of the comparatively narrow spectrum currently available.

Accordingly, a compromise which has been adopted at the ATCC, may be considered. The banking agency can determine the class, subclass and quantity of immunoglobulins secreted by a hybridoma and perform the other standard characterisations required (*Figure 1*). In addition, the isoelectric focusing (IEF) pattern of the immunoglobulin molecules produced can be established. This information should prove sufficient to monitor effectively for cross-contamination after the originator verifies monoclonal specificity from progeny of the initial seed stock.

4.5.1 *Applying the Ouchterlony Test for Immunoglobulin Isotypes*

The Ouchterlony method for determining the isotypes of immunoglobulins produced by hybridomas is relatively straightforward. *Table 13* gives the procedure for making the agar plates.

(i)   Remove a sample from each of the test hybridoma cultures which are close to maximum in cell density and collect the supernatants after centrifugation (200 *g* for 10 min).

(ii)   Place 35 μl of the supernatants in the outer wells. If the titre of antibody in the culture fluid is low, it may be necessary to repeat-load the well for that sample up to four times.

(iii)   Place 35 μl of the specific anti-mouse immunoglobulin typing antiserum in the centre well.

(iv)   Maintain the plates at room temperature in a humid environment overnight and read plates the next day. Precipitation lines should form between the well with the appropriate hybridoma supernatant and the specific known anti-immunoglobulin in the centre well. A typical example is shown in *Figure 9*.

4.5.2 *Quantitation of Secreted Immunoglobulin by Radial Immunodiffusion*

The amount of immunoglobulin produced by a hybridoma can be determined once the isotype is known. The appropriate typing antiserum is mixed with agar (*Table 14*). A known concentrate of immunoglobulin from the hybridoma supernatant is then added to an antigen well in the agar layer. As the antigen diffuses into the agar a ring of antigen-antibody precipitate forms around the well. The diameter of this ring is measured and compared with that of standards and a standard curve to permit quantita-

Table 13. Preparation of Plates for the Ouchterlony Test.

1.   Dissolve agarose to give 0.8% and sodium azide to make 0.025% in PBS by heating in boiling water bath.
2.   Add 10 drops of 0.5% aqueous trypan blue per 250 ml of solution and mix gently.
3.   Pour 1.5 − 2 ml aliquots of the hot solution into 60 × 15 mm Petri plates to form a thin bottom layer. While this is congealing retain the rest of the agarose solution in a 56°C water bath.
4.   After the primary layer has congealed add ~4 ml of the same agarose solution allow this to solidify, forming a secondary upper layer.
5.   Punch holes in concentric fashion and in the centre of the agar layer using a tubular well cutter, and aspirate the secondary upper layer of the plug to form each well with a pipette attached to any standard vacuum system.
6.   Store plates in the refrigerator until use.

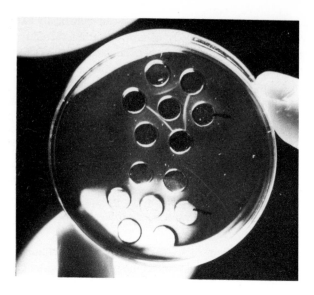

Figure 9. Assay for antibody isotype by the Ouchterlony technique. The precipitin bands between the centre and appropriate outer wells permit identification.

Table 14. Preparation of Diffusion Plates.

Buffer	In 100 ml DGDW dissolve 0.5 g of K_2HPO_4 and 0.6 g NaCl and adjust to pH 8.
Agar base	Add 1.5 g of Noble agar to 50 ml of buffer and dissolve by heating in a boiling water bath. Store in the refrigerator in $5-10$ ml aliquots.
Antiserum-agar mix	Dilute the typing rabbit anti-mouse antisera 1:10 in the pH 8 buffer and equilibrate in a 56°C water bath. Liquify agar base in a boiling water bath and equilibrate at 56°C. Mix equal volumes of the two solutions, dispense immediately onto 2 × 3 glass slides and allow to harden.

tion of the unknown amount placed in the antigen well. After preparation of the agar plates proceed to quantitate immunoglobulin as follows.

(i)   Collect 25 ml of the cell-free supernatant from the test hybridoma culture which is at or close to maximum cell density and cool to 4°C.

(ii)   Precipitate proteins by addition of 25 ml of a saturated solution of ammonium sulfate at 4°C, and centrifuge at 10 000 g for 30 min.

(iii)   Discard the supernatant and dissolve the precipitate in $1-3$ ml of distilled water noting the final volume and degree of concentration (typically 8- to 16-fold). Store at -60°C until ready for assay.

(iv)   Using a tubular well cutter, make 12 wells 3 mm in diameter and 12 mm apart in the agar on the diffusion plate(s). Aspirate out the plug with a Pastuer pipette attached to any standard vacuum system.

(v)   Place 10 μl of antibody concentrate into each test well.

(vi)   For the standard readings, place 10 μl of serial dilutions of the appropriate, purified mouse immunoglobulin into wells to include concentrations of 1, 0.5, 0.25 and 0.125 mg/ml.

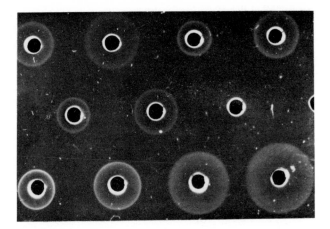

Figure 10. Quantitation by immunodiffusion of immunoglobulins produced by several hybridoma lines. The upper wells contained unknowns and the lower wells a series of standards ranging from 1.25 to 10 μg of immunoglobulin. (Photograph courtesy of W. Siegel.)

(vii)   Place the plates in a sealed, humid chamber at room temperature and leave for 16 − 18 h.

(viii) Measure the diameter of the precipitation rings, and calculate the immunoglobulin present by referring to the standards.

Many hybridomas produce 25 − 75 μg of immunoglobulin per ml of growth medium. A photograph showing results of a typical assay is presented as *Figure 10.*

4.5.3 *Isoelectric Focusing for Further Identity*

Since the majority of hybridomas are produced with murine fusion lines which express the same H-2 determinants, and many secrete immunoglobulin of the same idiotype, other means of identification are desireable. Isoelectric focusing (IEF) permits the resolution of as many as 50 000 mouse immunoglobulins. For this reason, the method is of use in further characterising hybridoma lines. Recipes for the agarose gel base and other solutions are given in *Table 15.*

To obtain satisfactory hybridoma supernatants for the assay:

(i)   Transfer hybridoma cells in approximate logarithmic growth to a medium consisting of RPMI 1640 (or other enriched chemically defined nutrient mixture) plus insulin (5 μg/ml), transferrin (5 μg/ml) and selenious acid sodium salt (5 μg/ml). The serum in standard media will interfere with the IEF assay.

(ii)   Wash the cell population in three changes of this medium, collecting the cells after each wash by cenrifugation at 200 *g* for 10 min. Resuspend without substantially reducing the cell-number-to-fluid-volume ratio.

(iii)   Maintain the hybridoma cells at 37°C for 5 days in this medium, centrifuge the cell suspension and collect the supernatant.

(iv)   Concentrate proteins in the fluid by precipitation with ammonium sulphate as in the immunodiffusion assay (Section 4.5.2). A minimum concentration of 1.7 − 2 mg/ml is required for IEF. It may be necessary to pass the concentrated

Table 15. Reagents for IEF of Hybridoma Immunoglobulins.

Agarose 1	1% (w/v) in DGDW. Dissolve by heating in boiling water bath and maintain at 56°C if for immediate use.
Pre-blended ampholyte	For pH range 3.5 – 9.5 (commercially available[a])
Fixative	150 ml methanol
	25 ml Trichloroacetic acid solution (100% w/v)
	86 ml sulphosalicylic acid (20%)
	350 ml DGDW
Collection fluid	0.01 M KCl
De-staining solution	700 ml ethanol
	200 ml acetic acid
	1100 ml DGDW
Staining solution	500 ml destaining solution
	2.5 g Coomassie blue. Filter before use
Catholyte	1.0 M NaOH
Anolyte	1.0 M H_3PO_4

[a]FMC Corporation, Rockland, Maine.

protein solution through a G-25 Sephadex column to remove traces of ammonium sulphate prior to IEF as this salt can interfere with the assay.

To perform the isoelectric focusing, agarose gels in a flat bed electrophoresis apparatus may be used (e.g., Pharmacia FBE 3000 with ECPS 3000/150 power supply).

(i)   Add 1.25 ml of ampholyte (2.5%) for a pH range of 3.5 – 9.5 to 20 ml of molten agarose at 56°C and mix gently.

(ii)   Using a pre-heated 25 ml pipette evenly place the agarose solution onto the hydrophilic surface of Gel-bond film (110 x 125 mm), and centre using the mould spacer.

(iii)   Allow the agarose to congeal, remove the mould and leave the gel in a humid chamber at 4°C for about 1 h. The manufacturer's booklet gives directions and figures to aid in pouring and placing the gel.

(iv)   Place the sample orientation grid on the platform using kerosene as an insulator, remove any trapped air, and circulate cooled water (5 – 10°C) through the platform.

(v)   Carefully blot off condensation from the agarose gel using filter paper and place the gel on the orientation grid.

(vi)   Trim two electrode wicks to 85 mm, soak each in the appropriate electrolyte and carefully centre them 5 mm from the anode or cathode ends, respectively.

(vii)   Position the applicator mask firmly on the gel 25 mm from the cathode, and apply standards (10 µl) and unknowns (15 µl with at least 25 µg protein) leaving the edge slots blank to allow for marginal drift.

(viii)   Place the electrophoresis lid over the electrode strips ensuring even contact and apply constant power (7 W) with voltage and milli-amps set at 3000 and 150, respectively.

(ix)   After 25 min turn off the power, remove any residual sample fluid, condensate and the applicator, and continue the run for another 60 min. Repeat the blotting to remove condensate as often as necessary during this period.

When the IEF run is complete turn off the power and proceed with gel sampling, fixation, staining and destaining as follows:

(i)   Blot the gel thoroughly, cut out 16 small discs along the anode-cathode gradient and collect these in numbered tubes, each containing 1 ml of 0.01 M KCl.

(ii)   Place the gel in a flat container, add fixative solution, leave for at least 10 min and rinse three times with DGDW.

(iii)   Mix the contents of each tube from step (i) above well and determine the pH within 4 − 24 h.

(iv)   Place blotting filter paper on the gel surface, invert over multiple layers of paper towelling and place a light weight on the Gel-bond backing.

(v)   After 20 − 30 min, remove the towelling, dry the gel under a hot air drier, re-wet the overlying filter paper with DGDW and gently peel it off of the gel.

(vi)   Dry the gel once more, stain it for 10 min in the Coomassie blue solution, rinse well with DGDW, and destain overnight in the ethanol/acetic acid solution.

(vii)   Dry the gel thoroughly, count the band numbers and note the locations with regard to the pH profile as determined from samples taken in step (iii).

A photograph of a typical resultant gel is given as *Figure 11*, and results on a wide variety of hybridomas are provided in reference (8). For further detail on methodology consult reference (48).

5. CELL SOURCE INFORMATION BANKS

A large number of lines are currently being stored in cell banks and private laboratories throughout the world. In many cases information concerning availability and special characteristics of these lines has not been widely communicated. This is especially true for hybridomas, both because of the importance of the new technology for their development and because so many potentially unique hybridomas can be generated with each fusion.

It would be difficult or impractical for any institution to bank, characterise, catalogue and distribute all of the cell lines and hybridomas available. Accordingly, groups of interested scientists and organisations have elected instead to develop computerised banks of relevant information on the myriad of lines available for use by the scientific and educational community. The role of such cell source information banks (CSIB) is generally that of clearing houses of information relating to the sources and unique characteristics of the specialised cell lines and hybridomas. No certification or endorsement of the cells is made. In most cases they are available directly from individual scientists and laboratories participating in the program.

CSI banks have originated comparatively recently. Their number and evolution with time may differ from the currently conceived plans as indicated below, depending upon funding, community input and community requirements. The ATCC initiated a broad range but marginally supported program of this type in 1977 and currently has source information on some 6000 metazoan cell lines and strains, many of which are catalogued. Other national cell banks such as The Human Genetic Mutant Cell Repository at The Institute for Medical Research in Camden, New Jersey and The Animal Cell Culture Collection being developed by the Public Health Laboratory Service, Porton Down,

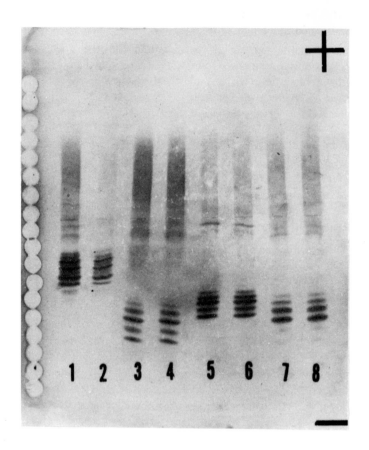

Figure 11. Isoelectric focusing patterns obtained from supernatant concentrates of four hybridoma lines. Samples were run in duplicate at $1-1.5$ mg/ml. **Lanes 1** and **2**, ATCC.HB2 (T3-A1); **lanes 3** and **4**, ATCC.HB5 (16-3-22S); **lanes 5** and **6**, ATCC.HB6 (17-3-3S); and **lanes 7** and **8**, ATCC.HB8 (19/178C$_1$). (Photograph courtesy of Y. Reid and A. Hamburger.)

Salisbury, UK also provide information both on the lines in their collections and on availability of specific cell cultures banked elsewhere.

5.1 CODATA-IUIS Hybridoma Data Bank

The Committee on Data for Science and Technology (CODATA) and the International Union of Immunological Societies (IUIS) are sponsoring the development of a computerised data bank on cloned cell lines and their immunoreactive products in cooperation with the World Health Organisation and the ATCC. Other scientific agencies which are currently participating include the Mission Interministerielle de l'Information Scientifique et Technique (MIDST) in France; the Institute for Physical and Chemical Research (RIKEN), Japan; The Fonds National Suisse de la Recherche Scientifique in Switzerland; the Medical Research Council in the United Kingdom; and the National Institutes of Health (NIAID, NIDR, NIGMS, NCI, DDR), Food and Drug Administration in the USA.

This data bank is intended for the use of the scientific community and is being planned and monitored by a CODATA Hybridoma Task Group consisting of nine immunologists from the various participating nations.

Input from the scientific community is required and is currently being solicited. Data including developer's name and address, pertinent references, if any, hybridoma designation, description of soluble product, detection procedures, availability of cells or supernatants and so forth are recorded by the originator on forms available from the ATCC. There is no minimum amount of information required for any given entry.

The program is non-profit making. When sufficient data have been entered into the system, the intent is to respond to queries without charge.

5.2 MIRDAB

A microbiological resource data bank (MIRDAB) has also been initiated by Excerpta Medica. The intent is to store and ultimately make available for a fee source and relevant scientific data on animal and plant cells, all forms of microorganisms, viruses, bacteriophages, plasmids, genes and gene vectors.

Separate forms for data entry on animal and plant cells and viruses have been generated and are now available from Elsevier Science Publishers. Information such as contributor's name and address, derivation histories, media or substrates required, literature references, secreted antibody or other product, availability and so forth is solicited.

6. ACKNOWLEDGEMENTS

The author wishes to acknowledge the assistance of many members of the Cell Culture Department in developing the various protocols referred to in this chapter. He is especially grateful to M. L. Macy and T. R. Chen for their suggestions and critical review of the manuscript. The work was supported under a grant (1-R26-CA25635) from the National Large Bowel Cancer Project of the National Cancer Institute, contract No 1-CB-14351 from the National Cancer Institute and contract (NO 1-RR-2-2139) from the Division of Research Resources of the National Institutes of Health.

7. REFERENCES

1. Schaeffer,W.I. (1979) *In Vitro*, **15**, 649.
2. Nelson-Rees,W.A. and Flandermeyer,R.R. (1977) *Science (Wash.)*, **195**, 1343.
3. Nelson-Rees,W., Daniels,W.W. and Flandermeyer,R.R. (1981) *Science (Wash.)*, **212**, 446.
4. Nelson-Rees,W.A., Flandermeyer,R.R. and Hawthorne,P.K. (1974) *Science (Wash.)*, **184**, 1093.
5. Barile,M.F. (1977) in *Cell Culture and Its Application*, Acton,R.T. and Lynn,J.D. (eds.), Academic Press, London and New York, p. 291.
6. McGarrity,G.J. (1982) *Adv. Cell Culture*, **2**, 99.
7. Hay,R.J., Williams,C.D., Macy,M.L. and Lavappa,K.S. (1982) *Am. Rev. Respir. Dis.*, **125**, 222.
8. Hay,R.J., Macy,M.L., Hamburger,A., Weinblatt,A. and Chen,T.R. (1983) *American Type Culture Collection Catalogue II*, 4th edn., Rockville, MD, p. 12,
9. Hay,R.J. (1984) in *Markers of Colonic Cell Differentiation*, Wolman,S.R. and Mastromarino,A.J. (eds.), Raven Press, New York, p. 3.
10. Stulberg,C.S. (1973) in *Contamination in Tissue Culture*, Fogh,J. (ed.), Academic Press, London and New York, p. 1.
11. Macy,M.L. (1978) *Tissue Culture Association Manual*, **4**, 833.

12. Hsu,T.C. and Benirschke,K. (1967 − 1975) *An Atlas of Mammalian Chromosomes*, 9 volumes, published by Springer-Verlag, New York, Heidelberg and Berlin.
13. Dilworth,S., Hay,R.J. and Daggett,P.-M. (1979) *Tissue Culture Association Manual*, **5**, 1107.
14. Freshney,R.I. (1983) *Culture of Animal Cells*, published by Alan R. Liss, Inc., New York.
15. Cour,I., Maxwell,G. and Hay,R.J. (1979) *Tissue Culture Association Manual*, **5**, 1157.
16. Macy,M.L. (1980) *Tissue Culture Association Manual*, **5**, 1151.
17. Rovozzo,G.C. and Burke,C.N. (1973) *A Manual of Basic Virological Techniques*, published by Prentice-Hall, Inc., Englewood Cliffs, New Jersey.
18. Hay,R.J., Kern,J. and Caputo,J. (1979) *Tissue Culture Association Manual*, **5**, 1127.
19. Povey,S., Hopkinson,D.A., Harris,H. and Franks,L.M. (1976) *Nature*, **264**, 60.
20. O'Brien,S.J., Kleiner,G., Olson,R. and Shannon,J.R. (1977) *Science (Wash.)*, **195**, 1345.
21. O'Brien,S.J., Shannon,J.E. and Gail,M.H. (1980) *In Vitro*, **16**, 119.
22. Wright,W.C., Daniels,W.P. and Fogh,J. (1981) *J. Natl. Cancer Inst.*, **66**, 239.
23. Rutzky,L.P. and Siciliano,M.J. (1982) *J. Natl. Cancer Inst.*, **68**, 81.
24. Halton,D.M., Peterson,W.D.,Jr. and Hukku,B. (1983) *In Vitro*, **19**, 16.
25. Nichols,E.A. and Ruddle,F.H. (1973) *J. Histochem. Cytochem.*, **21**, 1066.
26. Stoner,G.D., Katoh,Y., Foidart,J.-M., Trump,B.F., Steinert,P.M. and Harris,C.C. (1981) *In Vitro*, **17**, 577.
27. Pollack,M.S., Heagney,S.D., Livingston,P.O. and Fogh,J. (1981) *J. Natl. Cancer Inst.*, **6**, 1003.
28. Seabright,M. (1971) *Lancet*, **2**, 971.
29. Sun,N.C., Chu,E.H.Y. and Chang,C.C. (1973) *Mammalian Chromosome Newsletter*, **Jan.**, 26.
30. Chen,T.R., Hay,R.J. and Macy,M.L. (1982) *Cancer Genet. Cytogenet.*, **6**, 93.
31. Anderson,N.G. and Anderson,N.L. (1979) *Behring Inst. Mitt.*, **63**, 169.
32. Naylor,S.L. (1984) *In Vitro*, in press.
33. Luft,J.H. (1961) *J. Biophys. Biochem. Cytol.*, **9**, 409.
34. Rash,J.E. and Fambrough,D. (1973) *Dev. Biol.*, **30**, 166.
35. Ramaekers,F.C.S., Puts,J.J.G., Kant,A., Moesker,O., Jap,P.H.K. and Vooijs,G.P. (1982) *Cold Spring Harbor Symp. Quant. Biol.*, **46**, 331.
36. Schlegel,R., Banks-Schlegel,S., McLeod,J.A. and Pinkus,G.S. (1980) *Am. J. Pathol.*, **101**, 41.
37. Akeson,R. (1977) *J. Natl. Cancer Inst.*, **58**, 863.
38. Sasaki,M., Peterson,J.A. and Ceriani,R.L. (1981) *In Vitro*, **17**, 150.
39. Papsidero,L.D., Kuriyama,M., Wang,M.C., Horoszewicz,J., Leong,S.S., Valenzuela,L., Murphy,G.P. and Chu,T.M. (1981) *J. Natl. Cancer Inst.*, **66**, 37.
40. Koprowski,H., Steplewski,Z., Mitchell,K., Herlyn,M., Herlyn,D. and Fuhrer,P. (1979) *Somatic Cell Genet.*, **5**, 957.
41. Metzgar,R.S., Gaillard,M.T., Levine,S.J., Tuck,F.L., Bossen,E.H. and Borowitz,M.J. (1982) *Cancer Res.*, **42**, 601.
42. Kung,P.C., Goldstein,G., Reinherz,E.L. and Schlossman,S.F. (1979) *Science (Wash.)*, **206**, 347.
43. Yasumura,Y., Tashjian,A.H.,Jr. and Sato,G.H. (1966) *Science (Wash.)*, **154**, 1186.
44. Ambesi-Impiombato,F.S., Parks,L.A.M. and Coon,H.G. (1980) *Proc. Natl. Acad. Sci. USA*, **77**, 345.
45. Kimes,B.W. and Brandt,B.L. (1976) *Exp. Cell Res.*, **98**, 349.
46. Soule,H.D., Vazquez,J., Long,A., Albert,S. and Brennan,M. (1973) *J. Natl. Cancer Inst.*, **51**, 1409.
47. Wigley,C.B. (1975) *Differentiation*, **4**, 25.
48. Reid,Y., Breth,L. and Hamburger,A.W. (1984) *J. Tissue Cult. Methods*, in press.

CHAPTER 5

Separation of Viable Cells by Centrifugal Elutriation

1. INTRODUCTION

In many cell biology experimental studies, the investigation of cell populations consisting of a single cell-type confers great advantage over the inherently complicated examination of mixed cell populations. As an extension of this concept, studies relating cellular function to the cell division cycle have a requirement for a single cell-type population separated, and so synchronised, with respect to the position of the cells in the cell division cycle.

Between 1965 and 1970, the successful development of isopycnic (buoyant density) and velocity sedimentation methods, alone or in sequential combination, resulted in numerous reports of useful cell separations. However, prior to this period of general interest in cell separation, Lindahl in 1948 (1) described a centrifuge and cell separation chamber for the 'counter-streaming centrifugation' of cells. From 1955 this apparatus was used by Lindahl and colleagues to enrich specific cell types from mixed cell populations and pre-mitotic or mitotic cells from randomly proliferating ascites tumour populations (2). The technique was named centrifugal elutriation (3).

2. CELL SEPARATION METHODS

The theory of sedimentation as applied to cells has been reviewed in detail previously (4). Isopycnic or buoyant density sedimentation of cells in continuous gradients requires sufficient force and/or time for cells to relocate to respective density bands within the gradient. Since separation is effected on the basis of respective cell densities, and the densities of many cell types overlap broadly, the application of this method has some limitations (5). A further disadvantage is the high centrifugal force often used and the possible toxicity of the gradient medium employed. Nevertheless, in combination with sedimentation velocity, the method has provided highly enriched populations of viable cells. For example, 80 – 100% erythropoietic stem cells have been isolated from haemopoietic populations and separated into S phase and non-S phase fractions in Percoll (6).

Velocity sedimentation separates cells on the basis of both cell density and cell diameter. Sedimentation at unit gravity (7) is reliable and technically easy to operate and has the advantage of using inexpensive equipment such as the Staput apparatus (8) or the CelSep chamber (9) (Du Pont Ltd.).

The apparatus can accommodate a limited number of about 10^8 cells maximum layered onto a gradient of serum, bovine serum albumin or Ficoll. Problems arise at the sample/gradient interface at high cell number where cells stream down into the gradient. However, streaming can be minimised by using an intermediate layer between cells and gradient. A major disadvantage of unit gravity sedimentation is the long separation times required.

Sedimentation in an isokinetic gradient (e.g., Ficoll in tissue culture medium) also offers the advantage of reliable cell separation in sterile conditions using standard equipment normally available. As in the case of unit gravity sedimentation, relatively small cell numbers are processed.

The theory of centrifugal elutriation has been discussed by Sanderson *et al.* (10). At a low g force the system separates up to 10^9 cells in a very short time. Throughout the procedure the cells may be suspended in their normal growth medium in sterile conditions and, prior to loading the rotor, there is no requirement for pelleting the cells thus avoiding minor perturbations and maximising cell viability. The method requires more skill, and the equipment used is more expensive than for some other sedimentation procedures. Hydrodynamic shear stresses may cause problems although these are alleviated somewhat by elutriating in the presence of serum. Reaggregation of cells is a problem in most forms of velocity sedimentation, more so when separating a cell suspension derived from disaggregated tissue than when fractionating an established cell line. In centrifugal elutriation the formation of cell aggregates interferes with the flow of elutriation buffer. Nevertheless, as a high resolution separation device, centrifugal elutriation can produce synchronised populations of minimally perturbed viable cells.

Electronic sorting of cells labelled with monoclonal antibodies specifically binding to surface markers is an extremely effective alternative to the use of sedimentation methods (see Chapter 6). For example, Watt *et al.* (11) describe the differential binding of two monoclonal antibodies to haemopoietic cells. When used sequentially in conjunction with flow cytometry an enriched population of up to 60% committed erythroid precursors is obtained. Relative to centrifugal elutriation only small cell numbers are separated in a brief process time. The cell sorting apparatus is expensive to purchase and costly to maintain.

3. THE CENTRIFUGAL ELUTRIATOR

In 1973 Beckman marketed the JE-6 Elutriator Rotor and around 1980 this model was upgraded to the currently available JE-6B rotor which runs in a standard J2-21 centrifuge.

The complete apparatus for continuous flow elutriation consists of the JE-6B rotor containing the elutriation chamber, a pump and additional equipment for loading/elutriating the cell sample or collecting elutriated fractions.

3.1 **Elutriator Rotor**

The elutriation rotor comprises the following main parts.
(i) Black anodised aluminium body containing a central stainless steel column with entry and exit apertures.

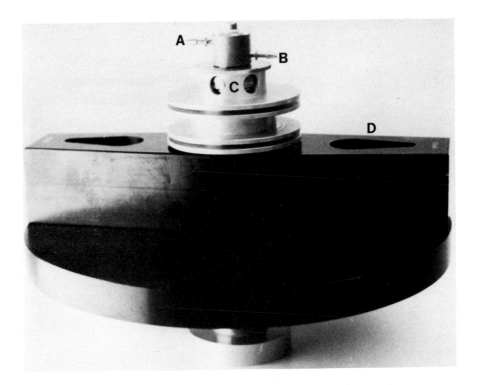

Figure 1. An assembled elutriator rotor showing the sample input (A) and output (B), the rotating seal (C) and the aperture (D) for viewing the elutriation chamber and contents by means of a stroboscopic light.

(ii) A separation chamber and counterbalanced bypass chamber connecting directly with the central column.
(iii) A rotating plastic seal which allows continuous entry and exit of the sample and elutriating liquid to the separation chamber while the rotor is spinning.

The assembled rotor is shown in *Figure 1* with the stationary input and output connections sited above the rotating seal. The aperture on the top and underside of the rotor body permits observation of the chamber within and its contents.

3.2 Elutriation Chamber

The standard separation chamber shown in *Figure 2A* is constructed from clear epoxy-resin. The sample, or elutriating buffer, enters from the side at the base of the chamber such that the flow of liquid directly opposes the centrifugal force. A modified J2-21 centrifuge lid incorporating a viewing port allows the spinning rotor chamber to be seen, illuminated by a synchronised stroboscopic light. The chamber filled with buffer appears as in *Figure 3* which also shows the transmitted image of the strobe lamp filaments.

During loading of a cell suspension, a cell boundary can be observed advancing gradually towards the centripetal exit from the chamber (*Figure 3*) assuming that the sedimentation rate of the cells in the spinning rotor is exactly balanced by the flow

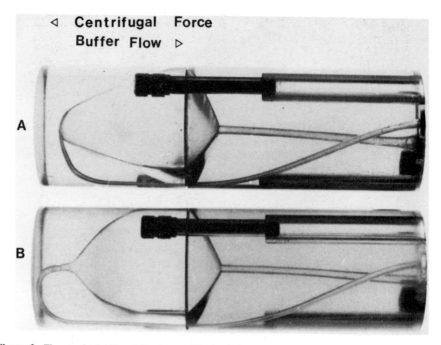

⊲ Centrifugal Force
Buffer Flow ⊳

A

B

Figure 2. The standard (**A**) and Sanderson (**B**) elutriation chamber.

rate of the fluid through the chamber. The discrete cell boundary transiently disappears when a fraction of cells is elutriated in response to a small increase in flow rate. Thereafter the boundary is reestablished and again visible as equilibrium is restored between the two opposing forces acting on the remaining cells.

The standard chamber's geometry, with continuous taper, produces a gradient of flow rates from one end to the other. Therefore, cells with a wide range of different sedimentation rates can be held within the chamber. As an alternative option, the Sanderson chamber (supplied by Beckman) can be used to retain cells of differing sedimentation rate in the lower part of the chamber where the walls diverge rapidly (*Figure 2B*). This reservoir of cells advances, as the flow rate is increased incrementally, into the upper chamber where the walls are almost parallel. This results in greater resolution, separating cells differing minimally in physical characteristics. No cell boundary is visible at the centripetal end of this chamber. A further contrast from the standard chamber is in the inlet at the base. In the Sanderson chamber, cells and elutrating fluid enter at the base directly from below rather than from the side. This helps to counteract any tendency for the cells to form clumps.

3.3 **Peristaltic Pump and Additional Apparatus**

A variable speed peristaltic pump regulated by a micrometer speed control is required to provide the flow of elutriating buffer. The pump should preferably have automatic compensation for changes in load torque demand and provide a linear relationship between the setting of the pump speed control potentiometer and the observed flow rate over the useful range of $10-60$ ml/min. Examples of such pumps are the Watson-

D. Conkie

Figure 3. The standard elutriation chamber as viewed *in situ* after loading the cell sample. Arrows indicate advancing cell boundary and inlet at base of chamber.

Marlow 501 and the Cole Parmer Masterflex 7520.

The pump draws the cell suspension or the elutriation buffer from containers at constant temperature in a water bath or ice bath. Each container is open to the atmosphere via a 0.22 μm filter (Millipore or Gelman), and is connected by silicone tubing (0.06 – 0.12 inch internal diameter and 0.06 inch wall thickness, Cole-Parmer) and Teflon valves as shown in *Figure 4*. The pump is also connected to the rotor input via a pressure gauge which indicates any line blockage, and a bubble trap which collects stray air bubbles preventing their entry to the chamber and disruption of the suspended cells. The output from the rotor passes via a flow cell (Starna) to the sterile

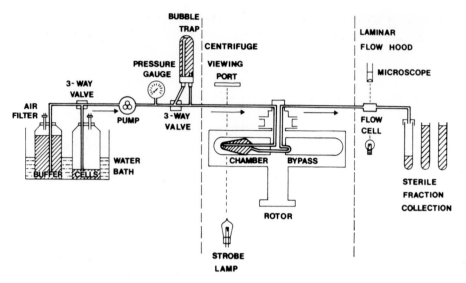

Figure 4. Diagramatic representation of the complete elutriation apparatus.

collection area within a laminar flow hood. Observation of the flow cell with a low power microscope or absorption at 660 nm permits monitoring of the fractions as they are collected and indicates, when free of cells, that a subsequent fraction can be elutriated. This also ensures that discrete fractions are collected in a minimum volume.

3.4 Complete Elutriation System

The detailed arrangement of the rotor and ancillary equipment is flexible and can be varied to suit individual experimental requirements. For example, it is possible to assemble the apparatus with the pump on the rotor output line. This tends to reduce the turbulence observed at the base of the separation chamber. In another variation the cell sample can be introduced between the pump and the input to the rotor. This avoids passage of the cells through the pump head. However, in practice, no deleterious effect arising from passage through the pump head is obvious as judged from cell viability studies.

When cells are loaded from an upstream reservoir as shown in *Figure 4*, tangentially acting Coriolis forces affect the initial low cell number in the chamber resulting in random elutriation of cells of all sizes (10). Therefore, as an alternative, a concentrated cell sample may be injected from a syringe into the elutriating medium immediately upstream of the rotor. This bulk sample arriving in the chamber results in an equilibrium between the inertial and hydrodynamic forces.

4. TECHNICAL PROTOCOL

4.1 Sterilisation of the Apparatus

Elutriation can be performed without sterilisation of the apparatus when no subsequent tissue culture of the separated cell fractions is required. Otherwise, the following procedures should be followed. Prior to assembly of the apparatus, autoclave sections of the tubing and connections to be used with the buffer and cell sample or for fraction collection.

The manufacturers recommend sterilisation of the complete assembled apparatus by pumping through 6% hydrogen peroxide. As alternatives, either 0.2% diethyl pyrocarbonate or Cidex (Arbrook, Livingstone, UK) are useful. Occasionally, the entire assembled rotor, with the exception of the separation chamber gasket, may be autoclaved at 121°C for about 1 h. The vinyl gasket can be separately sterilised in 6% hydrogen peroxide. Because of a possible fire hazard from damaged or sparking electrical connections resulting from a system failure, disinfection of the assembled apparatus by pumping through ethanol is inadvisable while the centrifuge circuits are live.

(i)   Assemble the apparatus as show in *Figure 4* but with the buffer and cell sample replaced by vessels each containing approximately 200 ml of Cidex.

(ii)   With the rotor stationary, start the pump and set the speed such that about 40 ml/min of the sterilising agent is drawn into the system. Vary the three-way valve settings so that all tubing and apparatus are filled.

(iii)   Invert the pressure gauge to remove air from the neck, then return to normal position.

(iv)   Similarly, invert the bubble trap and purge all air by temporarily withdrawing the small needle to the inner level of the cap. Reposition the bubble trap as in *Figure 4*.

(v)   To purge air from the rotor, start the centrifuge and accelerate the rotor to 1000 r.p.m. Switch off the pump and then switch off the rotor. Restart the pump when the rotor speed drops below 500 r.p.m. Air bubbles should flow out of the rotor. Repeat the procedure until no further air bubbles emerge from the rotor. The pressure gauge should read less than 10 p.s.i. (1.67 bar).

(vi)   Switch off the rotor and pump and leave the apparatus filled with Cidex overnight.

(vii)   After sterilisation, clamp both liquid input lines and transfer them with sterile precautions to vessels each containing 1000 ml sterile distilled water. Release the clamps.

(viii) Pump the water through the system, collecting Cidex for future use.

4.2 Flow Rate Calibration

(i)   Start the rotor and accelerate to 2000 r.p.m.

(ii)   When the rotor speed has stabilised, switch on the strobe control and adjust the delay until the elutriation chamber is 'stopped' and clearly visible through the viewing port (*Figure 3*).

(iii)   Obtain from the strobe counter an accurate record of the rotor speed.

(iv)   Set the pump potentiometer to a low arbitrary setting and measure the flow rate by collecting the effluent in a measuring cylinder over 2 min.

(v)   Repeat for incremental increases in the pump potentiometer and from the data construct a graph or program a microprocessor to indicate the relationship between the pump setting and flow rate.

4.3 Sample Loading

(i)   Using sterile technique, if required, clamp the input lines and remove them from the distilled water containers, taking care to avoid introduction of air.

(ii) Transfer the input tubes to containers of 1 l and 100 ml of elutriation buffer. As an example, buffer may be a balanced salt solution for elutriation of cell fractions when no further culture is required. Ideally, complete culture medium should be used to elutriate cells which are subsequently to be recultured. The medium may be maintained at 37°C in the water bath as shown (*Figure 4*).

(iii) Unclamp the tubes and use appropriate settings for the three-way valve to pump most of the 100 ml aliquot into the input tube. This will displace the water in the tube. Rearrange the three-way valve to pump buffer from the 1 l container through the system until all water is displaced. Retain the bulk of the 1 l sample as the elutriation buffer.

(iv) Clamp and remove the input tube from the 100 ml aliquot container. Replace in the cell sample, as in *Figure 4*, and release the clamp. A suitable cell sample may be 2×10^8 cells suspended in 200 ml of complete medium.

(v)   Set the pump flow rate to 12 ml/min and check that the rotor speed is 2000 r.p.m. Rearrange the three-way valve to pump the cell sample into the separation chamber. This condition will retain in the chamber all particles greater than 8 μm diameter. If peripheral red blood cells are loaded at this setting, they will immediately pass through the chamber to the collection vessel.

(vi) When about 95% of the cell sample has been loaded, adjust the three-way valve to bring the elutriation buffer on line once again. When subsequently all of the cells have been pumped from the bubble trap, adjust the local three-way valve to bypass the bubble trap.

4.4 Elutriation of Discrete Cell Populations

There are two methods of altering the equilibrium of the two opposing forces (sedimentation rate in the centrifugal field and the flow rate) acting on the cells suspended in the separation chamber. Either the centrifugal force is reduced or the flow rate is increased.

Although the modification to the standard speed control potentiometer on the J2-21 centrifuge is available to aid accurate rotor speed adjustment, the rotor deceleration tends to overshoot the set speed. Furthermore, the strobe control must then be readjusted to view the chamber at each new lower speed selected for discrete cell population elutriation. Thus, in practice, it is best to use constant rotor speed and variable flow rate.

(i)   By advancing the pump control, increase the flow rate of the elutriation buffer by about 5 ml/min. The exact increment required to elutriate a specific cell fraction is determined empirically in the first instance. However, some guidance is given from the expression:

$$F = XD^2 \left(\frac{RPM}{1000}\right)^2 \quad \text{where:}$$

F is flow rate at the pump in ml/min; X is 0.0511 (Standard chamber) or 0.0378 (Sanderson chamber); D is cell diameter in μm; and RPM is rotor speed.

This equation is an expression of conditions at the elutriation boundary and is used in the Beckman manual to construct a nomogram relating flow rate, cell size and rotor speed. The same data can be modified, for example, using a microprocessor program, to incorporate a correction factor obtained from the relationship between calculated and actual cell size of cells elutriated at any combination of rotor speed and flow rate.

(ii)   Observe the cell boundary advancing centripetally in the separation chamber and cells passing through to the flow cell. Using sterile conditions if required, collect a cell fraction. About 150 ml of elutriation buffer may be needed to collect all of the cell fraction.

(iii)   When the cell boundary is again visible in the separation chamber and the flow cell is free of cells, a further incremental increase in the pump flow rate will elutriate the next cell fraction for collection as before.

Elutriation of a specific cell fraction at a given flow rate and rotor speed is reproducible provided that the separation is performed using the same elutriation buffer, the same temperature and the same initial cell density in the chamber. Constant rotor speed and constant flow rate may be combined with step increases in elutriating medium density for high resolution fractionation. Note that the entire chamber contents will be pumped from the chamber if the rotor is switched off. This technique may be useful when concentrating a specific cell type in the chamber such as white blood cells from whole blood. Failure of the pump causes the cells in the chamber to pellet.

5. COLLECTION AND INTERPRETATION OF DATA

The information which can be obtained from elutriation experiments depends on whether the initial sample is a heterogeneous or single-cell type population.

5.1 Heterogeneous Cell Populations (Assessment of Cell Fraction Purity)

Normally the aim of an elutriation experiment in this case would be to separate and collect distinct cell types. For example, host cells may be removed from a tumour cell suspension. Diagnosis of the cell type(s) present in each fraction can be accomplished by classical cytological fixation and staining methods. Morphological studies indicate the degree of cellular integrity, particularly when mechanical or enzymatic disaggregation is used to produce the initial cell suspension from an organ or solid tumour (see Chapters 1 and 6). The enrichment of cell fractions can be further estimated by measurement of cell size with a Coulter particle size analyser, biochemical methods, clonal culture in soft agar or other *in vitro* colony forming assays and the ability of cell fractions to colonise *in vivo*.

An example of a simple separation of cell types from a mixture is the elutriation of avian erythrocytes and mouse erythroleukaemia cells. The erythrocytes are elutriated under the conditions used to load the mixed population (2000 r.p.m., flow rate of 14 ml/min), whereas the larger erythroleukaemia cells are retained in the chamber to be elutriated later at 2000 r.p.m. and 25 ml/min. The initial mixed population containing 55% red blood cells (RBCs) and 45% erythroleukaemia cells results in an elutriated fraction greater than 99% RBCs with a second fraction containing more than 97% erythroleukaemia cells.

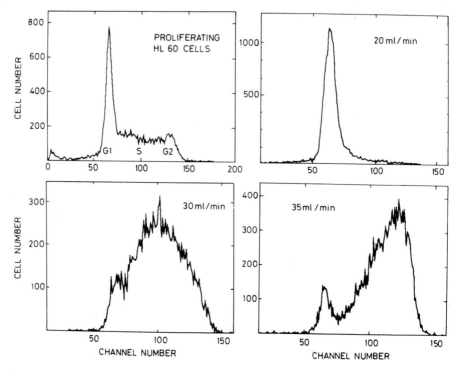

Figure 5. Flow cytometry profiles of HL60 cells randomly proliferating, together with separated fractions elutriated at the flow rates shown.

5.2 Established Cell Lines (Recovery and Viability)

Alternative procedures are required in the classification of cell fractions elutriated from established cell lines of homogeneous cell type. In this case centrifugal elutriation can yield fractions of cells separated on the basis of their position in the cell division cycle. Morphological criteria do not readily distinguish between the different cell cycle stages.

To distinguish 'S' phase cells from other cell cycle phases, the incorporation of tritiated thymidine into DNA can be used followed by a suitable assay for the isotope such as autoradiography or scintillation counting. However, by far the most informative analysis of elutriated fractions is obtained by flow cytometry (see Chapter 6).

Two examples are provided which demonstrate the resolution of the method and the viability of the separated cells.

5.2.1 *HL60 Cells*

The flow cytometry profile obtained for randomly dividing HL60 cells is shown in *Figure 5*. This profile is fairly typical for established cell lines even though HL60 cells have an inherent heterogeneity of DNA content. 2×10^8 of these cells are loaded at 15 ml/min with a rotor speed of 2000 r.p.m. The elutriation buffer is complete RPMI1640 medium containing 10% foetal bovine serum at 20°C. At a flow rate of 20 ml/min flow cytometry reveals a fraction of G1 cells. Both G1 cells and early

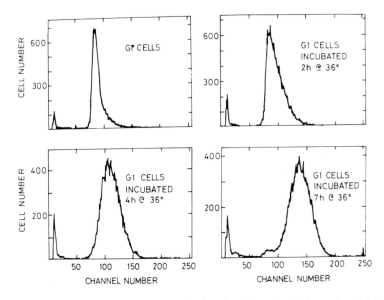

Figure 6. Flow cytometry profiles of G1 erythroleukaemia cells and G1 cells recultured for the times shown. The small peak in the first few channels represents avian erythrocytes added as a marker from which the distance to the G1 peak may be measured. Twice this distance marks the G2 position.

S phase cells are collected at 25 ml/min (not shown). A fairly pure mid-S phase fraction is collected at 30 ml/min and at 35 ml/min late S and G2 cells are obtained with a small contaminating peak of G1 cells. The total yield is typically greater than 80% leaving a proportion of aggregated cells from all cell cycle stages in the elutriation chamber.

5.2.2. *Erythroleukaemia Cells*

As a measure of viability, elutriated G1 erythroleukaemia cells may be recultured in elutriation buffer (complete medium) at 37.5°C. At various times during culture, samples are prepared for flow cytometry. The profiles (*Figure 6*) reveal a fairly synchronous progression through the cell division cycle obvious between 2 and 7 h and complete by 12 h (one cell cycle time) when the cells re-enter G1 (not shown).

Thus centrifugal elutriation is a procedure producing minimally perturbed synchronous cell populations at low *g* force, avoiding the deleterious effects of pelleting or the osmotic effects of density gradient centrifugation media. Furthermore, brief separation times are possible with up to 10^9 cells processed each time the chamber is loaded.

The applications of centrifugal elutriation are extensive and a bibliography is available from Beckman Ltd.

6. FUTURE DEVELOPMENT

Using the standard chamber the total yield of a specific cell type following elutriation of 10^9 cells may not always be adequate. Batch processing can increase total yield but also requires an increased time. A double separation chamber, where the bypass chamber is replaced by a second standard chamber in series, does not provide a significant advantage over two consecutive single chamber elutriations. The use of a double chamber system operating in parallel would require substantial modification to the rotor design. In principle, such a modified system need not be limited to two separation chambers.

However, the development of a scaled-up version of the original standard separation chamber has proved encouraging, maintaining the separation characteristics of the original system (12). The new high-capacity JE-10X Elutriator rotor marketed by Beckman has a 10-fold increased chamber volume. Preliminary reports suggest that separation characteristics and fraction purity are maintained with a 10-fold increase in yield.

7. ACKNOWLEDGEMENTS

Original work described is supported by grants from the Cancer Research Campaign.

8. REFERENCES

1. Lindahl,P.E. (1948) *Nature*, **161**, 648.
2. Lindahl,P.E. (1960) *Cancer Res.*, **20**, 841.
3. McEwan,C.R., Stallard,R.W. and Jukos,E.T. (1968) *Anal. Biochem.*, **23**, 369.
4. Pretlow,T.G., Weir,E.E. and Zettergren,J.G. (1975) *Int. Rev. Exp. Pathol.*, **14**, 91.
5. Pretlow,T.G. and Pretlow,T.P. (1982) in *Cell Separation*, Vol. **1**, Pretlow,T.G. and Pretlow,T.P. (eds.), Academic Press, London and NY, p. 41.
6. Nijhof,W. and Wierenga,P.K. (1982) *Exp. Haematol.*, **10**, Suppl. 12, 307.
7. Miller,R.G. and Phillips,R.A. (1969) *J. Cell Physiol.*, **73**, 191.
8. Miller,R.G. (1973) in *New Techniques in Biophysics and Cell Biology*, Vol. **1**, Pain,R.H. and Smith,B.J. (eds.), Wiley, NY, p. 87.
9. Wells,J.R. (1982) in *Cell Separation*, Vol. **1**, Pretlow,T.G. and Pretlow,T.P. (eds.), Academic Press, London and NY, p. 169.
10. Sanderson,R.J., Bird,K.E. Palmer,N.F. and Brenman,J. (1976) *Anal. Biochem.*, **71**, 615.
11. Watt,S.M., Metcalf,D., Gilmore,D.J., Stenning,G.M., Clark,M.R. and Waldmann,H. (1983) *Mol. Biol. Med.*, **1**, 95.
12. Jemionek,J.F. (1981) *Transfusion*, **21**, 268.

CHAPTER 6

Flow Cytometry

L. MORASCA and E. ERBA

1. INTRODUCTION

Flow cytometry is a relatively new technique that measures in each particle of a suspension several optical parameters such as optical density, fluorescence emission, and light scatter as single or multiple parameters. Each measurement is performed in times of the order of a few microseconds, permitting the collection of data on a large population of particles. This basic analytical function permits the further and more sophisticated function of sorting from a mixture of particles characterised by a well defined value of one or more of the parameters measured. Applied to cell biology this technique permits the characterisation of cell populations for light scattering, one parameter that is typical of this method of analysis, giving information on volume or surface area of the cells. The other two parameters usually evaluated are optical density and fluorescence emission.

Apart from cell sorting, a function more or less unique to this type of instrument, flow cytometry has significant advantages over conventional microfluorimetry and microdensitometry. Its main advantage is the ability to scan larger populations in a shorter time than conventional microtechniques, exposing fluorochromes to exciting light for a short and constant length of time. This gives more accurate and consistent measurements of emission. However, flow cytometry is unable to perform analysis of cellular structures and morphology of adhering cells while microfluorimetry and microdensitometry are specifically designed for these tasks.

Analysing cells by flow cytometry in an aqueous medium needs staining methodologies quite different from those needed with cells adhering to a slide: cells must be singly suspended, the stain or the fluorescent probe must be specific, must not diffuse out into the aqueous medium, must be excited by the few lines of an argon or krypton laser and if sorting of viable cells is required, the stain must be non-toxic.

In some simpler instruments an arc lamp substitutes for the costly laser, but, despite the larger number of spectral lines available, these arc illumination instruments are less sensitive. Computer size and programming is another factor limiting the application of flow cytometers. It must be stressed that the instrument configuration must be chosen carefully. Apart from cost, the choice of components will provide an instrument better suited to either certain types of analysis or to cell sorting. Some instruments may be excellent for multiparametric analysis but totally unsuited to cell sorting, and vice versa.

2. DESIGN OF EQUIPMENT

Instruments are based on the following constituent parts:

(a) A flow system capable of maintaining a constant flow of single cells in the excitation light focussed by a condenser lens.

(b) An optical system performing the following functions:

(i) production of a spot of excitation light a few microns in diameter in the area through which the cells are flowing;

(ii) measurement of the light loss produced by each cell passing through the light spot, the low angle forward scatter and eventually the spectral features of scattered light;

(iii) an optical system able to collect light scatter and the spectral features of fluorescence emitted by the cell at 90° to the exciting light path.

(c) A real time analogue system amplifying and analysing the signals.

(d) An analogue to digital converter able to memorise the flow of data in a multichannel analyser or in a computer.

(e) A programmable computer performing the analysis required.

(f) A source of light able to excite fluorescence in the single cell with an appropriate high energy density.

(g) In equipment designed to perform cell sorting the real time measurements can be selected by a threshold system able to drive a device charging the liquid stream containing the chosen cells by a positive or negative charge that will deflect each droplet to the right container as it is generated. It is therefore possible to separate from the flow two different types of cells by deflecting them to the positively or negatively charged area of collection.

Several companies are now involved in the production of these instruments and they fall into two main groups: analytical, able to measure scatter, fluorescence and absorption, and cell sorters able to separate cells with defined features.

Among analytical instruments the simplest ones are those using an arc lamp for excitation, as for example the ICP 22 and FACS Analyzer. More sophisticated instruments are more expensive and have the limited number of excitation lines produced by gas lasers, but provide the high energy density characteristic of these light sources and therefore elicit a more intense emission of fluorescence from the flurochromes in use.

Cell sorters all include one or two lasers for excitation and more sophisticated computer hardware. The design of instruments from different companies differs essentially in the flow chamber derived (a) from the need to have an efficient analytical system, rigid but easily operated as proposed by Ortho with model 3Q,50 and Spectrum, or (b) from the need to have more efficient sorting where the FACS series of instruments (Becton Dickinson) are probably the most widely known. Other companies such as Coulter and Partec, are now involved in construction and marketing of similar types of instrument. Each company is able to provide efficient instruments satisfying particular requirements, and in most cases it is the accurate definition of this requirement which will determine the choice of machine.

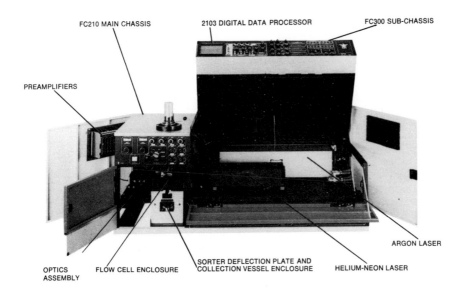

FC210 MAIN CHASSIS   2103 DIGITAL DATA PROCESSOR   FC300 SUB-CHASSIS

PREAMPLIFIERS

ARGON LASER

OPTICS ASSEMBLY   FLOW CELL ENCLOSURE   SORTER DEFLECTION PLATE AND COLLECTION VESSEL ENCLOSURE   HELIUM-NEON LASER

Figure 1. CYTOFLUORGRAF system 50H with access doors open.

2.1 Ortho Diagnostic Systems Inc ORTHO CYTOFLUOROGRAF* IIS

The CYTOFLUOROGRAF IIS (*Figures 1−3*) consists of four modules:

2.1.1 *Model FC210 Dual Laser Measurements Unit*

The FC210 contains two lasers, four detectors with their associated pre-amplifiers and gain controls, the flow cell, optical lenses and filters, and the sample delivery system.

Lasers. The standard configuration contains a 5 watt argon-ion laser and a 0.8 mW ultra-low noise helium-neon laser. The argon laser is used primarily for fluorescence excitation, and can also be used for light scatter measurements. It has continuously adjustable power output, and peak emissions at the following wavelengths: 488 nm, 514.5 nm, 501.7 nm, 457.9 nm, 496.5 nm, 418 nm, 351.1 nm.

The helium-neon laser is used for precise high resolution measurements of axial light extinction, narrow forward angle scatter and wide angle scatter. These capabilities permit measurement of cell size, use of non-fluorescent dyes and differentiation of cells based on differences in refractive index and granularity.

A 100 mW argon-ion laser may be substituted for the 5 W argon laser to reduce the cost.

Detectors. All detectors, except for the axial extinction which uses a solid state sensor, are ultra broad band, high sensitivity photomultiplier tubes (PMT). Four additional PMT sensors can be installed. Each PMT assembly contains a light emitting diode (LED) for calibration. A versatile optical filtering system permits selection of the wavelengths of light seen by each detector and the determination of angle of collection. Both fluorescence emissions and scattered light at the excitation wavelength can be collected.

*Trademark

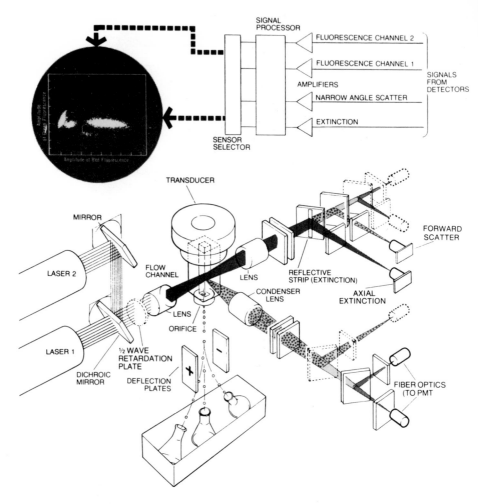

Figure 2. FC210 optics diagram. Key: diagonally placed elements are dichroic filters; orthogonally placed elements are filters based on wavelength, polarisation, etc.; dotted lines represent optional elements.

As a result both fluorescent and non-fluorescent dyes can be used in this system.

Altogether the optical system has seven detector positions; the use of fibre optic elements for coupling the optical signals to the photomultiplier detectors considerably enhances the versatility and ease of use of the system.

Flow Cell. The flow cell is constructed of quartz and has flat surfaces inside and out. The laser beams intercept the sample stream inside the flow cell (not in air).

Sample Delivery. The sample delivery system contains a volume sensing device for determining cell concentration, a disposable in-line filter to prevent clogging of the flow cell and a vacuum back flush. The sample entry chamber will accommodate up to a 30 ml beaker of cell suspension. All parts which come in contact with the measured sample can easily be removed for sterilisation.

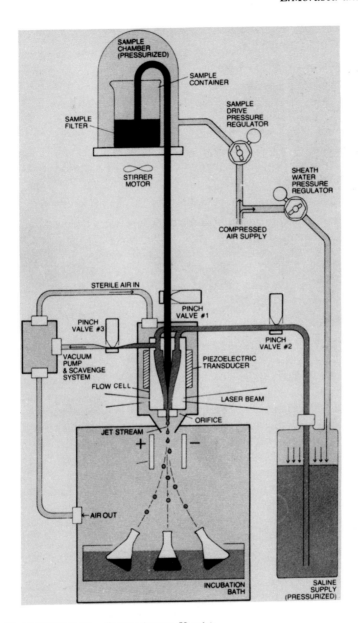

Figure 3. Model FC210 fluidics diagram (system 50 only).

2.1.2 *Model 2103 Digital Data Processor*

2103 is a multiplexed analyser which acquires, stores and displays two single parameter histograms of the data, displays cytograms (two parameter dot-plots) on a built-in storage oscilloscope and drives peripheral hard copy equipment.

2.1.3 *Model FC300 Multiparametric Signal Processor*

Among the many useful capabilities provided by FC300 are:

(a)   Selection of any of the four detector signals to be displayed on the X and the Y axes of the cytogram.

(b)   Selection of the detector signal mode of analysis (pulse height, pulse area or pulse width measurement).

(c)   Selection of logarithmic or linear scale on X and Y axes.

(d)   Determination of two regions of interest in the cytogram for differential counting and for sorting. These regions can be of any shape which is bounded by two pairs of parallel lines.

A sub-population for sorting can be defined by size and fluorescence, for example:

(e)   Setting of a delay to match the displacement between the two laser beams along the cell stream.

(f)   Observation of detector pulses as a function of time.

(g)   Provision on rear panel connectors of 45 or 120 microsecond held levels for all available signals accompanied by a strobe pulse, for convenient interfacing to a laboratory computer.

2.1.4 *Model FC400 Cell Sorter Module*

This module permits physical diversion of subpopulations of cells into separate collection vessels.

Coincidence rejection can be switched on or off. When on, this feature will allow collection of selected cells only if they are not accompanied by unwanted cells. This mode is used when high purity is required (95% or more). When switched off, the selected cells will be collected regardless of coincidences. This mode is used when maximum recovery of the unwanted cells is required as, for instance, where the wanted cells are rare.

A number of controls are provided on the front panel of the FC400. These include choice of the number of droplets for anti-coincidence detection independent of the number of droplets deflected per sort command, and selection of cells inside and outside the region of interest for collection. A time delay between signal generation and setting is provided.

A test mode is provided in which centre, right and left droplet streams are clearly visible. This mode is used to check the stability of the sorter.

The size and frequency of the droplets formed are determined mainly by the size of the orifice at the exit of the flow cell. Flow cells with orifices of three different diameters are available: 50 μm, 75 μm and 100 μm.

A preset sort control is provided. The instrument will then sort the required number of cells and stop automatically. An indicator light is on while the sort is in progress. A sort can be aborted and a continuous position is also provided which does not stop the sort automatically.

2.2 **Becton Dickinson Fluorescence Activated Cell Sorter, FACS 440** (*Figures 4 – 6*)

2.2.1 *Cell Detection*

As in the Ortho system, light scatter and fluorescence signals are derived from cells

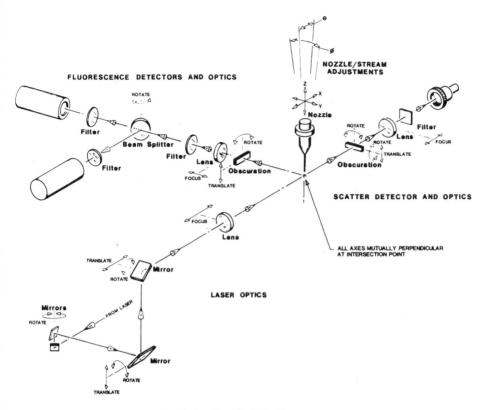

Figure 4. FACS 440. Sensor unit optical and mechanical adjustments.

as they traverse a high-intensity laser beam (*Figure 4*). They are illuminated by a 2 watt or a 5 watt u.v. argon-ion laser, having primary emissions of 800 mW at 514.5 nm and 700 mW at 488.0 nm and secondary emissions at 501.7 nm, 457.9 nm, 418 nm and 351.1 nm.

The major distinction between the FACS and the ORTHO CYTOFLUOROGRAF is that illumination of the sample takes place in the free liquid jet outside the cell alignment chamber as opposed to interrogation inside the flow chamber as in the CYTOFLUOROGRAF. Scattered light is detected by a solid state detector with spectral responses from 330 nm to 1100 nm. An iris-type aperture permits monitoring at narrow angles (1° to 1.5°) or wide angles (1° to 15°). The scatter detector is capable of resolving cells in the range from 0.3 μm to the limit of the nozzle aperture used.

(1° to 15°). The scatter dectector is capable of resolving cells in the range from 0.3 μm to the limit of the nozzle aperture used.

Fluorescence emission (350 – 600 nm) of prestained cells is detected by multiple photomultipliers. Specific dye emission is selected with the aid of multi-cavity band pass filters and a sharp transition dichroic mirror.

The fluorescence detector has an estimated lower limit of sensitivity of 2000 equivalent molecules of FITC per particle.

With the use of polarising filters as analysers in the fluorescence channel(s),

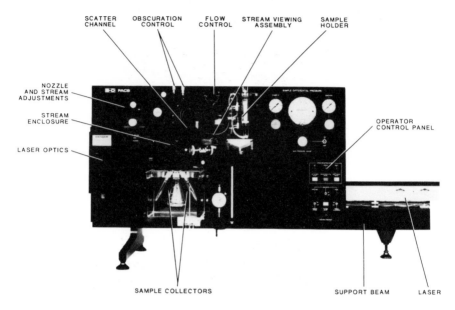

Figure 5. FACS 440 sensor unit.

measurements of horizontal and vertical polarisation components in fluorescence emissions can be made. The FACS 440 is also available with a Dual Laser Optical Bench. This option permits the simultaneous use of two laser beams for dual scatter and triple fluorescence excitation and sorting. Addition of the third fluorescence detection channel is required.

Using this dual laser set-up, 488 nm and 600 nm illumination is available for sequential excitation of fluorescein, phycoerythrin and Texas Red, as an example. The two independent beams intersect the cell stream at points vertically spaced so that a cell first crosses the 488 nm beam and then the 600 nm beam. Three fluorescence signals are generated for each cell, two from the first laser and one from the second laser. The signals from the two lasers are separated in space and time. The delay being the time required for the cell to travel from the first beam intersection point to the second beam intersection point. This time spacing permits the signals to be separately analysed by the electronics. The space separation allows the signal to be detected with optimum signal to noise ratio at every detector.

2.2.2 *Cell Sorting*

Sorting conditions are determined primarily by the selection of upper and lower thresholds for light scatter and fluorescence emission, independently or conjointly. There are two sample collection tubes (*Figure 5*) so that two specific sub-sets of cells can be sorted simultaneously. The unsorted residue, which can constitute a third sample, is collected in a central reservoir.

The sample chamber contains up to 5 ml of fluid, up to a cell concentration of 5×10^6 cells per ml. For samples of high concentration, where the instrument is used as a means

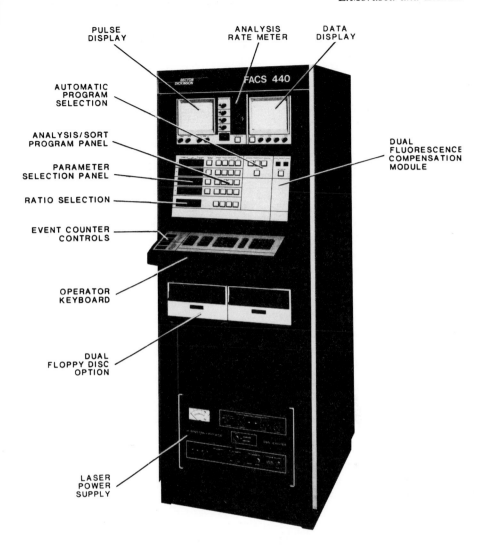

PULSE
DISPLAY

ANALYSIS
RATE METER

DATA
DISPLAY

AUTOMATIC
PROGRAM
SELECTION

ANALYSIS/SORT
PROGRAM PANEL

PARAMETER
SELECTION PANEL

RATIO SELECTION

EVENT COUNTER
CONTROLS

OPERATOR
KEYBOARD

DUAL
FLOPPY DISC
OPTION

LASER
POWER
SUPPLY

DUAL
FLUORESCENCE
COMPENSATION
MODULE

Figure 6. FACS 440. Electronics console.

of enrichment, virtually any concentration of cells can be used, provided the concentration of desired cells only does not exceed about 5×10^6 cells per ml, and the detected signals upon which the criterion of desirability of a specific cell is based exceed the background signal due to the unwanted cells.

Nozzles and filters are available permitting sorting with cells up to 30 μm in diameter. Cells up to 100 μm in diameter may be analysed routinely in the system.

After emergence the stream is intersected by the laser beam approximately 250 μm below the nozzle tip. By this design, the distance and time a cell takes in passing from the laser intercept to formation within a droplet is minimised. One result is maximum possible sorting rates while maintaining high purities. Also, since there is no intervening exit orifice between interrogation and sorting, any subsequent alteration of sample flow is avoided. Another important feature is that the possibility of flow cell contamination being in line with the incoming laser or light collecting optics is eliminated.

Droplets, generated by an electro-mechanical transducer, are charged by means of a positive or negative voltage pulse applied to the stream at breakoff. Deflection is achieved as the charged droplets pass between removable high-voltage electrodes. The deflection signals are synchronised with drop forming signals to provide the necessary precision.

Sorting rates up to 5000 cells per second can be obtained in the system. The sorting is limited by coincidence effects, and these are minimised by circuitry which aborts sorting when two cells are too close together to be separately isolated. When the desired cells are in very low abundance (i.e., 1 in 10^4) the abort circuitry can be disengaged, and higher sorting and analysis rates can be obtained. This provides an impure but enriched sample which may then be centrifuged and resorted with the abort circuitry engaged for final, high-purity separation.

Living cells maintain a very high level of viability after sorting, and the design of the FACS allows sterile collection to be undertaken.

The purity of the recovered fractions depends upon the concentration of cells, the rate of analysis, and the number of drops deflected. Purity ratios of 95% to 99% are possible. Sorting at the rate of 2000 cells per second and deflecting a single drop, recovery is approximately 85%. When deflecting two or more drops, recovery is approximately 90%. Results will vary depending upon sample preparation, delivery rate, orifice size and droplet rate.

2.2.3 *Signal Processing and Display*

The FACS 440 microcomputer provides both pulse height analysis and window selection functions. All control and selection entries are made using pushbuttons located on the control panel. The microcomputer is supplied with 4096 channels of memory as standard, and each channel has a capacity of 10^6 counts. Data are displayed on an X-Y oscilloscope as a dot plot or side by side histogram. Using direct plug-in interface boards, the FACS 440 can be interfaced with X-Y plotters, line printers and external computers (Consort series). Experimental data can be recorded as histograms via a RS 232 serial link or as raw data (list mode) via a parallel interface.

3. DATA OUTPUT

The optical signals produced by each cell flowing through the apparatus are detected by photomultipliers as analogue signals which are processed by some instruments, in real time, to derive the area of the signal, or its peak value, or its width at 0.5 of the peak value. The first derivations may permit the separation of cells with the same total fluorescence (same area signal) from cells with a different peak fluorescence signal.

The analogue output produces descriptive cytograms or introduces the signals to an analogue to digital convertor which transfers data to a multichannel analyser and stores it.

Computation may work on one, two or three parameters permitting correlation of data from one or more detectors on bidimensional or tridimensional histograms.

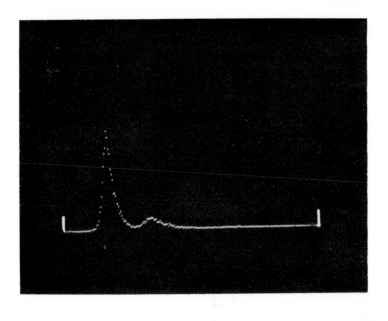

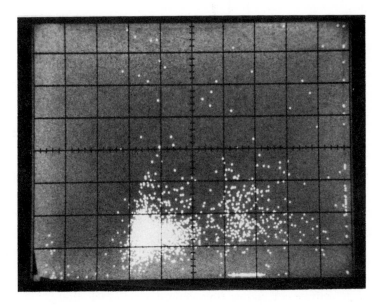

Figure 7. FHI cells (fetal human intestinal epithelium) stained with mithramycin as in text. Top figure cell number (ordinate) against fluorescence (abscissa). Bottom figure same cell sample shown as a 'dot plot' with light scatter on the ordinate and fluorescence on the abscissa (the gain on the fluorescence channel was increased from the top figure). Becton Dickinson FACS II. Data obtained with the assistance of B.D.Young and R.Sillar.

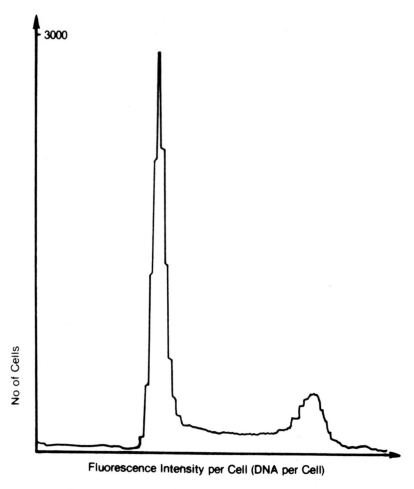

Figure 8. DNA staining with propidium iodide (from L.Morasca, Mario Negri Institute, Milan).

3.1 Descriptive Cytograms

Analogue signals produced by two photomultipliers (for example light scatter and fluor-escence) are collected in a storage oscilloscope and presented as a two dimensional array where each cell is represented by one dot (*Figure 7*, lower figure). Foci may then represent clusters of interest, and a threshold system can be activated to count particles or to activate cell sorting from defined areas of the display.

3.2 Histograms

These are produced by the multichannel analyser of the instrument plotting a single parameter, e.g., fluorescence against cell number (*Figure 8*), or as a three dimensional plot, e.g., fluorescence and light scatter against cell number. As the data are collected simultaneously from two different photomultipliers, this requires a more sophisticated multichannel analyser. A three parameters histogram can also be produced by some computers giving frequency distribution of cells for X, Y and Z parameters. Different configurations of instrument may be selected depending on the number of cellular parameters to be detected at the same time.

3.3 Computer Analysis of Data

The data collected on the multichannel analyser may be processed to obtain statistical data, to test fitting with theoretical models or to compare the results collected with others collected previously. These functions can be performed by different types of computer, some of which are provided as an integral part of the equipment and already carry some software designed to solve the most frequent types of analyses. Cheaper installations are possible provided the operator is prepared to do the programming.

4. PREPARATION OF SAMPLES

Flow cytometry measures one or more signals from each individual particle passing through the light path: therefore cells must be singly suspended and not aggregated, and no non-cellular debris must be present. The stain must be specific and not leach out into the carrying fluid.

4.1 Cell Suspension from Different Tissues

4.1.1 *Blood Samples*

Reagent and materials

Heparinised tubes
Hanks' BSS or Dulbecco's PBS
Ficoll-Hypaque or equivalent, 1.077 g/cc (Flow Laboratories, Pharmacia, Nygaard)
Universal containers (Sterilin) or 25 ml centrifuge tubes.

(i)   Blood cells do not need to be disaggregated but fibrin may form clots interfering with the flow system or aggregating single cells. Freshly drawn blood must therefore be collected into heparinised tubes.

(ii)   Dilute sample with 4 volumes of PBS or Hanks' BSS and layer 10 ml of diluted blood over 5 ml of Ficoll-Hypaque in a universal container or 25 ml centrifuge tube to eliminate red blood cells.

(iii)   Centrifuge at 1600 r.p.m. for 20 min.

(iv)   At the end of the centrifugation the mononucleate cells form a visible interface between the PBS and Ficoll. Aspirate the interface and resuspend in 10 ml PBS.

(v)   Centrifuge twice at 1200 r.p.m. for 10 min.

(vi)   Finally resuspend the pellet in Hanks' BSS at 10^6 cells/ml and analyse by flow cytometry.

4.1.2 *Disaggregation of Solid Tissues*

Enzymic disaggregation

The enzymes used for disaggregation of tissue may be divided into three classes: enzymes active on fibrous structures such as collagen or elastin (1), enzymes that hydrolyse mucopolysaccharides, e.g., hyaluronidase, and non-specific proteolytic enzymes, such as trypsin, pronase and dispase. Many factors must be taken into account using enzymatic disaggregation, such as the medium in which the enzyme is dissolved, duration of the exposure of the tissue to enzyme, temperature, pH and concentration of enzyme used and means by which the enzymatic activity is stopped (2).

(i)   *Trypsin.* A concentration of 0.25% crude trypsin (3) in PBS is effective with a wide variety of tissues $(4-9)$ with a $90-95\%$ viability even after prolonged exposure at 37°C (9). Trypsin is often used in combination with chelating agents (see below).

(ii) *Pronase.* Obtained from *Streptomyces griseus*, pronase is another enzyme widely used on a wide range of tissues $(10-14)$. There is no concrete evidence for the best concentration to be used for optimal tissue disaggregation (15) although a concentration of 0.1% of pronase is routinely employed in the technique described (12). Many data have shown that the effects of high concentrations of pronase for a short time are similar to those of low concentration for a relatively long period. Like trypsin, pronase has been used in combination with enzymatic agents such as DNase (14).

(iii) Collagenase. Used on many kinds of tissues such as mouse myeloma, mouse mast cell tumors, human Hodgkin's tumor, human spleen, human ovarian cancer (8,16,17).

Other proteases may be present in crude preparations of collagenase, while even purified preparations have been found to contain hyaluronidase (1). Purified collagenase may not yield a satisfactory single cell suspension, particularly with epithelial cells and may require supplementing with trysin or pronase. Collagenase may be used at different concentrations, from 100 to 1000 units/ml in culture medium for $15-45$ min at 37°C (8,17,18). Bacterial collagenase, like pronase, is not inhibited by serum, so prolonged digestion in complete medium with serum is possible, even up to several days' duration.

Chelating agents

Another approach used to obtain cell suspensions from solid tissue is to treat the tissue with chelating agents, usually in combination with enzymes. When used alone, chelating agents have been found to be less efficient than enzymatic treatment in producing intact cells. The most commonly used are ethylenediaminetetraacetic acid (EDTA), ethylene-glycol (2-aminoethylether)-n,nl-tetraacetic acid (EGTA) and citrate.

EDTA acts by complexing both Ca^{2+} and Mg^{2+} ions; it is used on different tissues at concentrations of $0.01-0.10$ M. EGTA is used at a concentration of 0.1 mM. Both are dissolved in PBS or BSS without Ca^{2+} and Mg^{2+}. Citrate is used at a concentration of 40 μM (19,20).

Mechanical disaggregation

Soft tissue containing small amounts of fibrous tissue can sometimes be disaggregated using mechanical disaggregation, although the number of viable cells obtained is always smaller than the number obtained using other methods. Another disadvantage of mechanical disaggregation is that many cells released from the tissue remain in clusters rather than fully dispersed as single cells. Mechanical disaggregation can be achieved by forcing fragments of tissue through sieves of progressively smaller size (e.g., 250 μm, 100 μm,

20 μm) or through hypodermic needles of progressively smaller size (e.g., 18G, 21G, 23G) (2,21,22).

It is difficult to recommend one single method for disaggregation of cells from solid tissues (especially for solid tumors) as one method may be used with good results, in terms of amount of debris and type of damage sustained by cells, for one particular tissue, but may not be suitable for another tissue. The following protocol is offered as a starting point. Further development may be necessary based on experience and reports in the literature.

Reagents and materials (sterile)

Scissors or pair of scalpels
Magnetic stirrer
Nylon or stainless steel gauze, 100 μm, 20 μm
Hanks' BSS
Enzyme solutions: 0.25% trypsin in PBS or PBS with 1 mM EDTA; 0.1% pronase (Sigma) in PBS or complete culture medium; DNase 100 μg/ml ($1-5$ μg/ml final concentration) (Sigma, Worthington). Collagenase crude CLS grade (Worthington) or grade 1 (Sigma), 2000 u/ml stock, $100-1000$ u/ml final.

(i) Place tissues in Hanks' BSS at 4°C. Mince the tissue into 1 mm³ pieces with scissors or crossed scalpels and wash twice with Hanks' BSS.
(ii) Remove Hanks' BSS and replace with enzyme solution for disaggregation at a ratio of $10-15$ ml:1 g tissue.
(iii) At the end of digestion, the enzyme activity is stopped by placing on ice or by addition of serum and the cells, decanted from the tissue fragments, centrifuged at 1200 r.p.m. for 10 min. Filtration through nylon or steel gauze, 100 μm followed by 20 μm, will help to remove aggregated cells and larger debris at this stage.

Note:

Never add DNase to these preparations if you plan to evaluate DNA content in the cells. Otherwise $1-5$ μg/ml may be used to minimise reaggregation due to DNA released from lysed cells.

4.1.3 *Fixation of cell suspension*

A cell suspension can be used immediately after preparation as living intact cells, or it can be fixed if staining and flow cytometry are to be carried out later, or if fixation is required for effective staining. Ethanol fixation is used most frequently.

Reagents and materials

Hanks' BSS or PBS Ca^{2+} and Mg^{2+} free with 1 mM EDTA
95% ethanol.

(i) Bring cells to a concentration of about 1×10^6 cells/ml in suspension in cold Hanks' BSS Ca^{2+} and Mg^{2+} free containing 0.5 mM EDTA.
(ii) Keeping the suspension at 4°C add 95% ethanol dropwise with constant mixing, in the ratio of three volumes for each volume of cell suspension, to give a final EtOH concentration of 70%.

(iii)   Cells can be kept in this suspension for a few months.

(iv)   Before using for cytophotometry, cells must be freed of EtOH by centrifugation and resuspended in PBS or whatever reagent is to be used for the procedure.

5. EXAMPLES OF TYPES OF ASSAY

5.1 Light Scatter of Unstained Cells

The presence of a particle, for example a cell, in the light beam, perturbs the beam and the incident light is re-distributed. The energy is re-emitted by the cells with a complex pattern where diffraction, refraction and reflection interact together in a function of the volume, surface configuration, and internal structures of the cell. Commercial instruments normally measure light scatter in two fixed positions, a small angle forward scatter, about 10°, and a 90° scatter.

The 90° scatter is considered to be affected mostly by diffraction due to the internal structures of the cell, while the low angle forward scatter is considered as more representative of cell volume (23 – 33) (*Figure 9*).

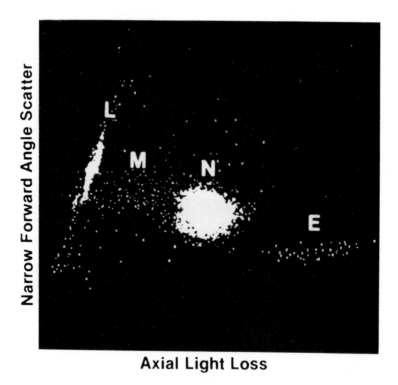

Figure 9. Light scatter of unstained cells (from Ortho Instruments, 410 University Avenue, Westwood, MA 02090). Differential counting of neutrophils, eosinophils, lymphocytes and monocytes in human blood by non-fluorescent method.

Scatter measurements

Reagents and materials

Hanks' BSS

(i)   Suspend cells in Hanks' BSS to obtain a concentration between 1×10^5 and 1×10^6 cells/ml.

(ii)   Fill the reservoir of the instrument containing the sheath fluid with the same balanced salt solution as used for the cells.

(iii) Set the flow of cells through the instrument to 500 cells/sec to obtain the best performance.

5.2 Staining Procedures

5.2.1 Suspension of Nuclei and DNA Staining with Propidium Iodide

Propidium iodide is an analogue of ethidium bromide and, like ethidium bromide, intercalates in the nucleic acid helix with a resultant increase in fluorescence of about 20-fold. Propidium iodide is about 1.8 times more fluorescent than eithidium bromide. At 488 nm excitation the emission maximum for the DNA propidium iodide complex occurs at approximately 615 nm (21,34 – 42).

A typical histogram is shown in *Figure 8* where the major peak corresponds to the G_1 phase of the cell cycle, the minor peak G_2/M and the intervening saddle S phase. Absolute values for DNA content can be obtained by using a standard such as chicken erythrocytes, run simultaneously with the sample.

DNA content of Lewis lung carcinoma of mouse (3LL)

Reagents

HBSS
Staining solution: 100 ml of sodium citrate 0.1%, 5 mg of propidium iodide, 900 μl of Nonidet P40 1% in H_2O.

(i)   Remove tumors from C57BL/6 mice and wash in a Petri dish with PBS.

(ii)   Remove obvious connective tissue and fat and mince with scissors.

(iii) Force the small pieces through a 18G $\times$ 1½ in. (1.20 $\times$ 38 mm) needle and resuspend in Hanks' BSS maintained at 4°C.

(iv)   Stain 3LL cell suspension by adding 3 ml of propidium iodide solution to 200 – 300 μl of cell suspension (500 000 cells/ml).

(v)   Store at 4°C for 20 – 30 min before analysis.

(vi)   Measure fluorescence emission between 580 nm and 750 nm to exclude the overlapping region of excitation and emission spectra of unbound propidium iodide.

5.2.2 Application to Cells in Culture

Reagents: as for 5.2.1.

(i)   Remove medium from culture dish and rinse twice with HBSS.

(ii)   Add 5 ml of propidium iodide solution and refrigerate the dishes for 10 min at 4°C.

(iii) Dislodge the nuclei by repeating pipetting and analyse by flow cytometry.

5.2.3 DNA Staining with Propidium Iodide in Whole Cells

Reagents

70% EtOH

Propidium iodide, 50 μg/ml, containing 40 μg/ml RNase

(i)   Fix cells in 70% EtOH (see 4.1.3).

(ii)   After centrifugation remove fixative, stain an aliquot of cells ($1 \times 10^5 - 1 \times 10^6$ cells/ml) with propidium iodide for 30 min at room temperature.

5.2.4 DNA Staining by Mithramycin and Propidium Iodide

In the combination staining technique using the fluorescent antibiotic mithramycin and propidium iodide the energy transfer occurs from the mithramycin donor molecule to the propidium iodide acceptor molecule.

     This technique is useful for staining and analysing solid tumor samples, sperm cells and gynaecologic material, as well as tissue culture cells (43).

Reagents

10 μg/ml mithramycin
10 μg/ml propidium iodide
15 mM magnesium chloride
0.1% Nonidet P40
Tris-HCl pH 7.5

(i)   The cell suspension can be used immediately after disaggregation or up to 1 week after fixation in 70% EtOH.

(ii)   Stain with propidium iodide/mithramycin solution for 1 h at 4°C.

(iii)   After staining run the sample through the instrument excited by a mercury 100 W lamp, with barrier filter BG12 3 mm plus one step filter K590.

5.2.5 DNA and RNA Staining by Acridine Orange

The metachromatic property of acridine orange is used for the differential identification of DNA and RNA. Acridine orange has been used as fluorescent stain for nucleic acids in fixed and unfixed cells, or as a marker of lysosomes, or in observing the metachromatic changes of fluorescence in dead cells, or to differentiate between cycling and quiescent cell populations.

     Despite the technical difficulties in obtaining a reproducible measurement of DNA and RNA, this method has been extensively used by some groups (44 – 57).

Differential staining of DNA and RNA

Reagents

Solution A: Stable for about 2 weeks, must be kept cold.
0.1 ml Triton X-100 (0.1%)
8 ml 1 N HCl
15 ml 1 M NaCl
76 ml distilled H_2O
100 ml total volume, pH 1.5

Solution B: Stable for several months.

10 ml 0.01 M EDTA
15.5 ml 1 M NaCl
31.5 ml 0.4 M Na_2HPO_4
18.5 ml 0.2 M citric acid
24 ml distilled H_2O
99 ml total volume, pH 6.0

Acridine orange stock solution

1 mg/ml in distilled H_2O.
N.B. Acridine orange is carcinogenic; handle with care.

Acridine orange working solution

0.1 ml of stock solution plus 9.9 ml of solution B.
Note: the sheath flow system of the instrument must be kept cold (4°C).
Argon laser excitation: 488 nm.
Red fluorescence is DNA (F >600 nm).
Green fluorescence is RNA (F >530 nm).

(i)   Add 0.2 ml of cells (8×10^6 cells/ml) suspended in PBS with 15% serum to
        0.4 ml of solution A and keep at 4°C for 45−60 sec.
(ii)   Add 1.2 ml of solution B for 2 min at room temperature.
(iii)   Analyse by flow cytometry within 10 min from the addition of solution B.

Note: Every step of the procedure is critical for obtaining reproducible results. It is
particularly important to have a constant number of cells to be stained; depending on
the ratio between nucleic acids and acridine orange, RNA metachromasia will be
affected. Time of exposure to staining solution and temperature will also strongly affect the reaction.

5.3 Application of Staining Procedures

5.3.1 *Differential Staining of Denatured and Double-stranded DNA*

Reagents

Hanks' BSS containing 1000 u/ml RNase A 0.2 M KCl, pH 1.35
Acridine orange, 5 μg/ml (16.7 μm) in 0.1 M citric acid, 0.2 M Na_2PO_4 buffer, pH 2.6.

(i)   Suspend 2×10^6 cells in 1 ml Hanks' BSS/RNase.
(ii)   Incubate at 37°C for 1 h.
(iii)   Mix 0.2 ml of cell suspension containing 4×10^5 cells (still in Hanks' BSS/RNase)
        with 0.5 ml of 0.2 M KCl pH 1.35 at 20°C for 30 sec.
(iv)   Stain by adding 2 ml acridine orange solution for 2 min.
(v)   The green (530 nm) and red (600 nm) fluorescence represent, respectively, the
        amount of denatured and double-stranded DNA of the cells.

Note: See note to 5.2.5.

5.3.2 *Cell Cycle Analysis by Bromodeoxyuridine (BrdU) and Hoechst 33258*

Hoechst 33258 stain fluoresces between 410 nm and 480 nm and needs to be excited
in the u.v. at 315 nm. It binds specifically to adenine-thymidine base pairs and is

therefore not only specific for DNA but also for the presence of thymidine. In the analysis of the cell cycle, the incubation of cells in the presence of BrdU substitutes BrdU for thymidine in DNA. Cells in the synthetic phase will therefore remain apparently diploid and after division will give rise to a new peak in the histogram of half the G_1 peak.

This technique permits direct measurement of the G_2 plus mitosis time to be made (58 − 63).

Reagents

Medium containing BrdU at 33 μg/ml and deoxycytidine at 26.4 μg/ml (equimolar) Staining solution: Hoechst 33258 in PBS. Cells are run through the instrument about 2 h after staining; stability of the stain is reported to range from 30 min to 24 h.

(i)   Add BrdU medium to the growth medium 1:10.
(ii)   Incubate cells for different lengths of time appropriate to the predicted cell cycle time.
(iii)   After incubation dissociate cells as for subcultivation.
(iv)   Resuspend the cells directly in the staining solution.

5.3.3 *DNA Staining of Viable Cells by Hoechst 33342*

Hoechst 33342 can be used as vital stain for DNA. It is known to produce relatively good stoichiometric DNA staining while preserving cell viability. Excitation is in the u.v. range of 350 − 363 nm, while emission is recovered over 450 nm (64 − 67).

Reagent

Hoechst 33342, 0.25 M in distilled H_2O

(i)   Prepare cell suspension at 10^6/ml and stain with Hoeschst 33342 at the A concentration of 5 − 10 μM for 20 min at room temperature.
(ii)   Do not wash the cells before analysis.

5.3.4 *Protein Staining by Fluorescein (FITC)*

Reagents

70% EtOH

Fluorescein isothiocyanate: 0.1 − 1.0 μg/ml in IBS containing 40 μg/ml RNase.

(i)   Fix cells in 70% EtOH for at least 18 h before staining.
(ii)   Centrifuge fixed cells to remove fixative.
(iii)   Stain cells for protein at room temperature for 30 min in fluorescein isothiocyanate solution.
(iv)   Analyse on flow cytometer with argon laser excitation at 488 nm and fluorescence emission 515 − 555 nm.

5.3.5 *Double Staining for DNA and Proteins*

For two colour staining, fixed cells are stained with a solution containing 18 μg/ml propidium iodide in 0.1% sodium citrate, 0.05 μg/ml FITC and 40 μg/ml RNase in PBS for at least 20 min at room temperature (*Figure 10*).

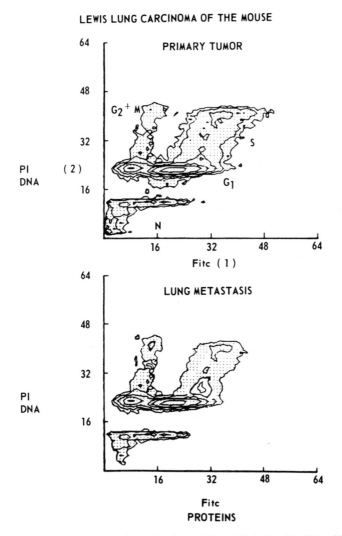

Figure 10. Double staining for DNA and proteins (from E.Erba, Mario Negri Institute, Milan).

5.4 Fluorescence Antibody Technique using Fluorescein

Direct immunofluorescence staining of cell surface

This technique is used to detect cells having specific membrane antigens by treating a cell population with monoclonal antibodies conjugated with fluorescein or rhodamine.

Reagents

FITC-conjugated antibody diluted in phosphate-buffered Eagle's MEM (PBMEM), containing 0.1% sodium azide and 2% calf serum. To test for appropriate concentration add 50 μl of different dilutions to cells in a microtitration plate and check for optimal staining by fluorescence microscopy. In analysis with anti-human LEU-1, dilute to

5 μg/ml or 0.25 μg/50 μl. Mouse or human cells at a concentration of 2 × 10^7 must be 90% viable.

(i)   Add 50 μl of the cells to a microtitration plate containing 50 μl diluted antibody per well and mix.
(ii) Incubate for 45 min on ice.
(iii) Centrifuge the plate at 500 g.
(iv) Discard the supernatant and wash the cell pellet twice with 100 μl of PBMEM. Centrifuge cells after each washing at 500 g for 3 min and aspirate supernatant from the pellets.
(v)   For flow cytometry analysis resuspend the cells in 1 ml of cold medium PBMEM to have a concentration of 1 × 10^6 cells/ml; keep cold before analysis (73,74).

6. DISCUSSION

Flow cyotometry provides a large set of instruments, similar but not identical, optimised for specific situations. Limiting factors are the sources of excitation light, the sensitivity of photomultipliers and the computerised functions available.

High resolution is always easily obtained with coherent light from low intensity lasers in a narrow beam of light focussed on the flow of cells.

High sensitivity can only be achieved with very costly high power lasers requiring the availability of a very high power supply (a multiline 18 watt output argon laser absorbs almost 50 KW/h and the 4 watt commonly used argon laser requires more than 12 KW/h). High excitation power is required, for example, when immunofluorescence of a small number of membrane receptors must be detected but also when fluorochromes require excitation in the u.v., where both argon and krypton lasers have low efficiency. Again, if a wide spectrum of wavelengths is required with a dye laser included in the system, initial costs and power required increases. To these apparently negative comments it must, however, be added that fluorochromes like propidium iodide permit analysis if nucleic acid content in vertebrate cells with the power given by a very small 100 mW laser, and that the mithramycin propidium iodide method works perfectly with a 100 W mercury lamp.

The performance of cell sorting must be considered separately from the analytical use of the instrumentation. This is characterised by the need of more sophisticated equipment from the electronic point of view but dedicated to a preparative function and not expected to perform at the limits of sensitivity and resolution.

Cell sorting provides cell suspensions of a good grade of purity in relation to the discriminating parameter chosen. All the methods described above can be applied in sorting. It must be noted, however, that only a few of them can be applied to living cells. This area requires further development. Further limitations in cell sorting that must be discussed are of technical origin: first of all cells are sorted in drops of fluid interspaced with about a 15-fold excess of empty droplets of the flow sheath. This means that sorted cells are strongly diluted and normally must be concentrated by further manipulation producing an unavoidable cell loss.

Speed of analysis greatly reduces the purity of preparations and the 'official' speed given by commercial firms of 5000 cells/sec must be considered a maximum giving fair quality preparations. A more efficient speed of analysis could be considered to be less than 2000 cells/sec. This speed of analysis applies to all the cells passing through the nozzle and the real recovery of sorted cells depends on the ratio between cells to be sorted and cells which are discarded due to the fact that the computer was still elaborating in real time information about previous cells. In practice a low frequency of droplets containing cells, 1:15 empty droplets, is required to avoid this problem and reduce the statistical chance of more than one cell occupying a droplet.

The time of sorting to select a sufficient number of cells of the desired type may therefore become very long, increasing the risk of poor performance linked to the probability of having in the cell suspension debris, aggregates, any type of foreign material able to clog the nozzle, or eventually modify the flux of fluid through the nozzle. This last possibility is the most serious in terms of cell loss and purity of the sorted cells prepared.

All these drawbacks of the methodology are presented in the light of our own experience. The advantages of flow cytometry and cell sorting are well known to most scientists and enthusiasm for the new possibilities offered by this technique tends to underestimate the difficulties and the limitations of the technology or of the type of equipment available.

7. ACKNOWLEDGEMENTS

Details of ORTHO CYTOFLUOROGRAF Flow Cytometer and Fluorescence Activated Cell Sorter FACS 440 reproduced by kind permission of Ortho Diagnostics Systems Inc. and Becton Dickinson (UK) Ltd, respectively.

8. REFERENCES

1. Waymouth,C. (1974) *In Vitro*, **19**, 97.
2. Pretlow,T.G., Weir,E.E. and Zettergren,J.C. (1975) *International Review of Experimental Pathology*, Vol. **14**, Richter,G.W. and Epstein,M.A. (eds.), Academic Press, New York, p. 91.
3. Mahler,H.R. and Cordes,E.H. (1966) *Biological Chemistry*, published by Harper & Row, New York.
4. Brattain,M.G., Kimball,P.M., Pretlow,T.G.,II, *et al.* (1977) *Br. J. Cancer*, **35**, 850.
5. Kreisberg,J.I., Pitts,A.M. and Pretlow,T.G.,II (1977) *Am. J. Pathol.*, **86**, 591.
6. Willson,J.K.V., Luberoff,D.E., Pitts,A., *et al.* (1975) *Immunology*, **28**, 161.
7. Dow,S.H. and Pretlow,T.G.,II (1975) *J. Natl. Cancer Inst.*, **54**, 147.
8. Morasca,L., Erba,E., Vaghi,M., Ghelardoni,C., Mangioni,C., Sessa,C., Landoni,F. and Garattini,S. (1983) *Br. J. Cancer*, **48**, 61.
9. Morasca,L., Erba,E., Vaghi,M., Amato,G., Pepe,S., Mangioni,C., Colombo,N., Landoni,F. and D'Incalci,M. (1985) *J. Exp. Clin. Cancer Res.*, in press.
10. Helmes,S.R., Brazeal,F.I., Bueschen,A.J., *et al.* (1975) *Am. J. Pathol.*, **80**, 79.
11. Pretlow,T.G.,II, Dow,S.R., Murad,T.M., *et al.* (1974) *Am. J. Pathol.*, **76**, 95.
12. Pretlow,T.G.,II, Jones,J. and Dow,S. (1974) *Am. J. Pathol.*, **74**, 275.
13. Pretlow,TG.,II, Scalise,M.M. and Weir,E.E. (1974) *Am. J. Pathol.*, **74**, 83.
14. Bowman,P. and McLaren,A. (1970) *J. Embryol. Exp. Morphol.*, **23**, 163.
15. Bowman,P. and McLaren,A. (1970) *J. Embryol. Exp. Morphol.*, **24**, 331.
16. Lasfargues,E.Y. and Moore,D.H. (1971) *In Vitro*, **7**, 21.
17. Kraehenbuhl,J.P. (1977) *J. Cell. Biol.*, **72**, 390.
18. Bashor,M.M. (1979) in *Methods of Enzymology*, Vol. **58**, Jakoby,W.B. and Pastan,I.H. (eds.), Academic Press, New York, p. 119.
19. Ham.R.G. and McKeehan,W.L. (1979) in *Methods in Enzymology*, Vol. **58**, Jakoby,W.B. and Pastan,I.H. (eds.), Academic Press, New York, p. 44.
20. Jacob,S.T. and Bhargava,P.M. (1972) *Exp. Cell. Res.*, **27**, 453.
21. D'Incalci,M., Torti,L., Damia,G., Erba,E., Morasca,L. and Garattini,S. (1983) *Cancer Res.*, **43**, 5674.
22. Erba,E., Ubezio,P., Colombo,T., Borggini,M., Torti,L., Vaghi,M., D'Incalci,M. and Morasca,L. (1983) *Cancer Chemothr. Pharmacol.*, **10**, 208.

23. Mullaney,P.F. and Dean,P.N. (1969) *Appl. Opt.*, **8**, 2361.
24. Mullaney,P.F., Crowell,J.M. and Salzman,G.C. (1976) *J. Histochem. Cytochem.*, **24**, 298.
25. Jovin,T.M., Morris,S.J. and Striker,G. (1976) *J. Histochem. Cytochem.*, **24**, 269.
26. Meyer,R.A. and Brunsting,A. (1975) *Biophys. J.*, **15**, 191.
27. Steinkamp,J.A. and Crissman,H.A. (1974) *J. Histochem. Cytochem.*, **22**, 616.
28. Sharpless,T., Traganos,F., Darzynkiewicz,Z. (1975) *Acta Cytol.*, **19**, 577.
29. Salzman,G.C., Growell,J.M., Martin,J.C., *et al.* (1975) *Acta Cytol.*, **19**, 374.
30. Price,B.J., Salzman,G.C., Crowell,J.M., *et al.* (1976) *Biophys. J.*, **16**, 64a.
31. Leary,J.F., Notter,M.F.D. and Todd,P. (1976) *J. Histochem. Cytochem.*, **24**, 1249.
32. Gledhill,B.L., Lake,S., Steinmetz,L.L., *et al.* (1976) *J. Cell. Physiol.*, **87**, 367.
33. Mullaney,P.F. and Fiel,R.J. (1976) *Appl. Opt.*, **15**, 310.
34. Erba,E., Vaghi,M., Pepe,S., Amato,G., Bistolfi,M., Ubezio,P., Mangioni,C., Landoni,F. and Morasca,L. (1985) *Br. J. Cancer*, **52**, 565.
35. Crissman,H.A., Mullaney,P.F. and Steinkamp,J.A. (1975) in *Methods in Cell Biology*, Vol. **9**, Prescott,D.M. (ed.), Academic Press, New York, p. 179.
36. Cram,L.S., Gomez,E.R., Thoen,C.O., *et al.* (1976) *J. Histochem. Cytochem.*, **24**, 383.
37. Krishan,A. (1975) *J. Cell Biol.*, **66**, 188.
38. Fried,J., Perez,A.G. and Clarkson,B.D. (1976) *J. Cell Biol.*, **71**, 172.
39. Bichel,P., Frederiksen,P., Kjaer,T., *et al.* (1977) in *Pulse Cytophotometry*, Lutz,D. (ed.), European Press, Ghent.
40. Barlogie,B., Spitzer,G., Hart,J.S., *et al.* (1976) *Blood*, **48**, 245.
41. Zante,J., Schuman,J., Barlogie,B., *et al.* (1976) in *Pulse Cytophotometry*, Gohde,W., Schuman,J. and Bucher,T.H. (ed.), European Press, Ghent, p. 97.
42. Ward,D.C., Reich,E. and Goldberg,I.H. (1965) *Science*, **149**, 1259.
43. Starace,G., Badaracco,G., Greco,C., Sacchi,A. and Zuppi,G. (1982) *Eur. J. Cancer Clin. Oncol.*, **18**, 973.
44. Bertalanffy,F.D. (1962) *Ann. N.Y. Acad. Sci.*, **93**, 715.
45. Nash,D. and Plaut,W. (1964) *Proc. Natl. Acad. Sci. USA*, **51**, 713.
46. Darzynkiewicz,Z., Traganos,F., Sharpless,T., *et al.* (1974) *Biochem. Biophys. Res. Commun.*, **59**, 392.
47. Darzynkiewicz,Z., Traganos,F., Sharpless,T., *et al.* (1975) *Exp. Cell. Res.*, **90**, 411.
48. Darzynkiewicz,Z., Traganos,F., Sharpless,T., *et al.* (1975) in *Pulse Cytofluorometry*, Haanen,C.A.M., Hillen,H.F.P., and Wessels, . (eds.), Ghent, European Press, p. 88.
49. Darzynkiewicz,Z., Traganos,F., Sharpless,T., *et al.* (1975) *Exp. Cell. Res.*, **95**, 143.
50. Darzynkiewicz,Z., Traganos,F., Sharpless,T., *et al.* (1976) *Proc. Natl. Acad. Sci. USA*, **73**, 2881.
51. Traganos,F., Darzynkiewicz,Z., Sharpless,T., *et al.* (1977) *J. Histochem. Cytochem.*, **25**, 46.
52. Darzynkiewicz,Z., Traganos,F. and Sharpless,F. (1976) *J. Cell. Biol.*, **68**, 1.
53. Traganos,F., Darzynkiewicz,Z., Sharpless,T., *et al.* (1975) *J. Histochem. Cytochem.*, **23**, 431.
54. Melamed,M.R. (1972) *Eur. J. Cancer*, **8**, 287.
55. Kapuscinski,J., Darzynkiewicz,Z. and Melamed,M.R. (1982) *Cytometry*, **2**, 201.
56. Traganos,F., Darzynkiewicz,Z. and Melamed,M.R. (1982) *Cytometry*, **2**, 212.
57. Darzynkiewicz,Z., Traganos,F., Xue,S., Coico,L. and Melamed,M.R. (1981) *Cytometry*, **1**, 279.
58. Beck.H.P. (1982) *Cytometry*, **2**, 170.
59. Weissblum,B. and Haenssler,E. (1974) *Chromosoma*, **44**, 255.
60. Steel,G.C., Adams,K. and Barret,J.C. (1977) *Growth Kinetics of Tumors*, published by Clarendon Press, Oxford.
61. Latt,S.A. (1975) *Chromosoma*, **52**, 297.
62. Heiber,L., Beck,H.P. and Huhle,C. (1981) *Cytometry*, **2**, 175.
63. Bohmer,R.M. (1979) *Cell Tissue Kinet.*, **12**, 101.
64. Loken,M.R. (1980) *Cytometry*, **1**, 136.
65. Hamori,E., Arndt-Jovin,D., Grmwade,B.G. and Jovin,T.M. (1980) *Cytometry*, **1**, 132.
66. Arndt-Jovin,D.J. and Jovin,T.M. (1977) *J. Histochem. Cytochem.*, **25**, 585.
67. Fried,J., Doblin,J., Takamoto,S., Perez,A., Hansen,H. and Clarkson,B. (1983) *Cytometry*, **3**, 42.
68. Crissman,H.A., Oka,M.S. and Steinkamp,J.A. (1976) *J. Histochem. Cytochem.*, **24**, 64.
69. Crissman,H.A. and Steinkamp,J.A. (1973) *J. Cell Biol.*, **59**, 766.
70. Shapiro,H.M. (1983) *Cytometry*, **3**, 227.
71. Crissman,H.A. and Steinkamp,J.A. (1982) *Cytometry*, **3**, 84.
72. Crissman,H.A., Egmond,J.V., Holdrinet,R.F., Pennings,A. and Haanen,C. (1981) *Cytometry*, **2**, 59.
73. Zeile,G. (1980) *Cytometry*, **1**, 37.
74. Colotta,F., Peri,G., Villa,A. and Mantovani,A. (1984) *J. Immunol.*, **132**, 936.

CHAPTER 7

Organ Culture

I. LASNITZKI

1. INTRODUCTION

Organ culture means the explantation and growth *in vitro* of organs or part of organs in which the various tissue components, such as parenchyma and stroma, and their anatomical relationship and function are preserved in culture, so that the explanted tissue closely resembles its parent tissue *in vivo*. The outgrowth of isolated cells from the periphery of the explants is discouraged and minimised by suitable culture conditions and the 'new' growth is composed of differentiated structures. Thus, in glands new glandular structures are formed, or in lung tissue new small bronchi develop at the periphery of the explant. They consist of alveoli lined with secretory, cuboidal or columnar glandular or bronchial epithelium.

In tissues, lined with squamous epithelium, such as skin or oesophagus, or in bladder lined with transitional epithelium, the epithelium follows a similar pattern of differentiation as in the organs *in vivo*. Hormone-dependent tissues remain hormone sensitive and responsive and endocrine organs continue to secrete specific hormones. Finally, in foetal tissues, morphogenesis *in vitro* closely resembles that seen *in vivo*.

2. SURVEY OF ORGAN CULTURE TECHNIQUES

Loeb (1) was the first to culture fragments of adult rabbit liver, kidney, thyroid and ovary on small plasma clots inside a test tube and found that they retained their normal histological structure for 3 days. His work antedates the introduction of tissue culture by Harrison (2) by 10 years.

Loeb and Fleisher (3) showed that the tube must be filled with oxygen to prevent central necrosis of the explants. Parker (4) also emphasised the necessity for oxygen and grew fragments of several organs in a shallow layer of medium in a flat-bottomed flask filled with 80% oxygen. This method proved unsatisfactory as most tissues sank to the bottom of the flask except skin which floated on the surface of the medium. Skin is not wettable and this property was used by Medawar (5) to grow slices of rabbit ear skin on a serum-saline mixture in a flask filled with 70% oxygen. However, a disadvantage of growing skin in fluid medium is that the fragments tend to curl up and the epithelium migrates and covers the dermal surface.

It had been recognised earlier that, with the exception of skin, most organ rudiments or organs could be better maintained growing on a solid support than in fluid medium.

2.1 Clotted Plasma Substrate

Fell and Robison (6) introduced the 'watchglass technique' by which organ rudiments or organs were grown on the surface of a clot consisting of chick plasma and chick embryo extract, contained in a watchglass. This became the classic standard technique for morphogenetic studies of embryonic organ rudiments. The method has later been modified to investigate the action of hormones, vitamins and carcinogens in adult mammalian tissues (7) and will be described in detail in Section 3.

Another type of culture vessel consisted of an embryological watchglass containing a plasma clot and closed with a glass lid sealed on with paraffin wax. This was first introduced by Rudnick (8) and later adopted by Gaillard (9). He used a clot consisting of two parts of human plasma, one part of human placental serum and one part of human baby brain extract mixed with six parts of a saline solution.

The plasma clot, although it supported the growth and development of foetal and adult organs, had several disadvantages. It usually became liquified in the neighbourhood of the explants so that they came to lie in a pool of medium. Moreover, because of the complexity of the medium no biochemical investigation was possible.

2.2 Agar Substrate

The problems encountered using plasma clots could be eliminated by the use of agar clots. The agar clot technique was first introduced by Spratt (10,11). Wolff and Haffen (12) modified Gaillard's technique and used an agar gel contained in an embryological watchglass. The agar method has been successfully used for developmental and morphogenetic studies and will, like the watchglass technique, be described in detail in Section 3.

Although the agar does not liquefy, it cannot be added to or analysed without transplanting the cultures. This disadvantage was overcome by the use of fluid media combined with a support which prevented the cultures being immersed.

2.3 Raft Methods

Chen (13) found that lens paper used for cleaning microscope lenses (Gurr, London) is non-wettable and will float on fluid medium. He explanted $4-5$ cultures on a 25×25 mm raft of lens paper which floated on serum in a watchglass. Richter (14) improved on this by treating the lens paper with silicone which prevented it sinking into the medium. Lash *et al.* (15) combined the lens paper with Millipore filters. They punched a small hole in the centre of the lens paper raft and covered this with a strip of Millipore filter. Different types of tissues were cultured on either side of the filter and their interaction with each other studied. Shaffer (16) replaced the lens paper with rayon acetate. The rayon acetate strips were made to float on the fluid medium by treating the four corners with silicone. The rayon acetate has the advantage over lens paper that it is acetone soluble and can be dissolved during the histological procedures by immersing it in acetone.

2.4 Grid Method

The use of rafts floating on a fluid medium did not provide ideal conditions. The rafts often sank and the tissues became frequently, and to different depths, immersed into the medium. This difficulty was overcome by the 'Grid' technique devised by Trowell (17). He introduced metal grids, made at first of tantalum wire gauze. This was replaced later by the more rigid expanded metal, obtainable as a continuous sheet of stainless steel (18), or by titanium (19). The grids were square with a surface of 25 × 25 mm, with the edges bent over to form four legs, and about 4 mm high. Skeletal tissues could be cultured directly on the grid, but softer tissues such as glands or skin were first explanted on strips of lens paper and these deposited on the grids. The grids with their explants were placed in a culture chamber filled with medium up to the level of the grid. The original Trowell technique was aimed at maintaining adult mammalian tissues which have a higher requirement for oxygen than foetal organs. To achieve this, the culture chambers were enclosed in containers which were perfused with a mixture of CO_2 and oxygen. The method succeeded in preserving the viability and histological structure of the adult tissues, such as prostate glands, kidney, thyroid and pituitary. The technique, particularly the application of the gas phase, has since been simplified and in this modified form has been and still is widely used.

2.5 Intermittent Exposure to Medium and Gas Phase

More recently, a method which provides intermittent exposure to medium and gas phase has been successfully used for the long-term culture of human adult tissues, including bronchial and mammary epithelium, oesophagus and uterine endocervix (20 − 23). By this technqiue, the explants are attached to the bottom of a plastic culture dish and covered with medium. The dishes are enclosed in an atmosphere controlled chamber which is filled with an appropriate gas mixture. The chamber is placed on a rocker platform and rocked at several cycles per minute during cultivation.

3. WATCHGLASS TECHNIQUE

The technique was introduced by Fell and Robison (6), originally, to study the development of avian limb bone rudiments but was extended to investigate growth and differentiation of other avian and mammalian tissues. By this method, avian tissues are placed on a clot consisting of chick plasma and chick embryo extract in equal proportions which is contained in a watchglass. One or two such watchglasses are enclosed in a Petri dish carpeted with moist cotton wool or filter paper (*Figure 1*) to prevent evaporation of the clot and explants. The Petri dishes are transferred to an incubator and incubated, usually at 37.5°C.

3.1 Preparation of Embryo Extract

3.1.1 *Materials*

All reagents and instruments must be sterile.
   Fertilised hen eggs incubated for 7 − 8 days
   Petri dishes
   Straight forceps
   Curved forceps
   Curved scissors
   Homogeniser, loose fitting, Potter-Elvejheim, glass with Teflon pestle

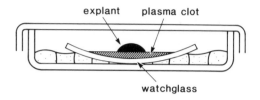

Figure 1. Watchglass technique, Fell and Robison, 1929 (6).

Centrifuge tubes, 50 ml (Corning)
Pasteur pipettes
Wide-mouthed pipettes made by cutting off the tops of ordinary pipettes
Storage vials
Hanks' solution (HBSS)
70% Alcohol.

3.1.2 *Protocol*

(i)   Before removing the embryos, wipe the blunt end of the eggs with cotton wool soaked in 70% alcohol to sterilise the surface of the shell. Remove the upper end of the shell carefully, by tapping the shell to crack it and peeling outwards, and tear open the inner shell membrane covering the embryo, with a pair of straight forceps. Lift embryos out by their heads using a pair of curved forceps and drop them gently into a Petri dish. Wash them thoroughly with HBSS, remove the eyes and mince the embryos by hand with a pair of curved scissors.

(ii) Transfer the minced tissue with a wide-mouthed pipette to a homogeniser and grind by hand for a few minutes.

(iii) Mix the resultant homogenate with an equal amount of HBSS and centrifuge the mixture at 20 *g* for 10 min at 4°C. Remove the supernatant and if necessary re-centrifuge to remove suspended cell debris. The final supernatant should be free of cells and appear opalescent.

(iv) The extract can be used immediately or transferred to vials in 1 ml aliquots for storage at −20°C. Use within 3 months.

3.2 Preparation of Chicken Plasma

Commercial preparations of plasma are not recommended as they usually do not produce a firm clot and may not support the same degree of growth and differentiation as freshly prepared plasma.

3.2.1 *Materials*

Domestic chicken 6−12 months old
Syringe, 30 ml hypodermic needles
Centrifuge tubes
Heparin.

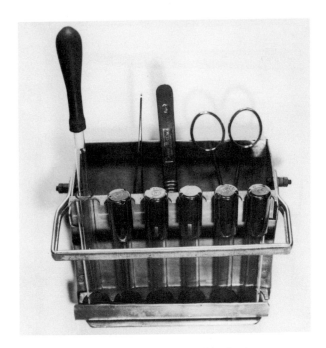

Figure 2. Stainless steel rack to hold Pasteur pipettes and dissecting instruments.

3.2.2 *Protocol*

(i)   Prepare chicken plasma by the standard technique described by Paul (24). To lower the lipid content of the plasma starve the animals for at least 24 h before bleeding.

(ii) To prevent premature clotting of the plasma, siliconise all glassware and keep cold, rinse the collecting syringe with a solution of heparin (0.01 − 1%). The amount of heparin should be kept as low as possible to permit clotting of the plasma when mixed with embryo extract.

(iii) Test plasma for sterility (see Chapter 4) and store in siliconised tubes at −20°C. Use within 6 months.

3.3 **Explantation**

Materials

Watchglasses 3.5 mm in diameter
Petri dishes 91 mm in diameter
Cotton wool or filter paper to fit Petri dishes
Sterile distilled water or saline solution
Graduated pipettes 10 ml
Pasteur pipettes
Wide-mouthed pipettes prepared by cutting off tops of Pasteur pipettes
Extra fine pipettes prepared by drawing out Pasteur pipettes to a fine point in flame
Pipette rack (*Figure 2*)

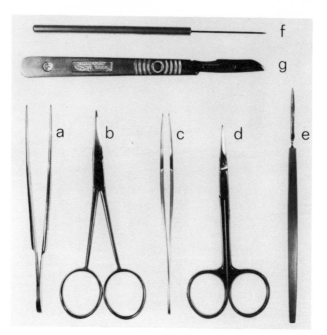

Figure 3. Dissecting instruments: **(a)** medium size straight forceps; **(b)** medium size straight scissors; **(c)** watchmaker's forceps; **(d)** smaller fine scissors; **(e)** cataract knife; **(f)** dissecting needle; **(g)** Swann-Morton knife.

Cavity slides 6−7 mm deep
Glass slides approximately 8 cm²
Chicken plasma
Chick embryo extract (50%)
Heat-inactivated (56°C for 30 min) sera; rat, horse, calf, for use with mammalian tissues
Glass rods for mixing plasma and extract
HBSS
Cataract knives, Swann-Morton blades, No. 11 ⎫
Dissecting needles ⎪
Watchmaker's forceps, straight forceps 10 cm ⎬ (*Figure 3*)
Medium size scissors ⎪
Fine scissors for dissection. ⎭

3.3.2 *Protocol*

(i) Cover the bottom of Petri dishes with a layer of cotton wool or, preferably, three layers of filter paper which have one or two 3.5 cm holes punched into them, and sterilise by dry heat (see Chapter 1).

(ii) On the morning of explanatation spread 10 ml of sterile distilled water or HBSS with a graduated pipette on cotton wool or filter paper and place the watchglasses over the punched holes. The holes will allow transmission of light for a dissecting microscope and facilitate macroscopic observations of the explanted tissues during cultivation.

(iii) For the cultivation of avian tissues, prepare clots by mixing 10 drops of chicken plasma and 10 drops of chick embryo extract (6). Stir the mixture quickly and thoroughly with a glass rod.

(iv) For mammalian tissues, heat-inactivated, rat, horse or calf serum is added to chicken plasma and chick embryo extract, usually in a proportion of chicken plasma: serum:extract of 2:1:1, but this may be modified according to the particular tissue used and the problem to be investigated. In experiments in which the effect of chemical agents on growth and differentiation is studied the compounds to be investigated are incorporated into the mixture before clotting (25).

(v) Using fine scissors and watchmaker's forceps, remove organs or part of organs, rinse them with HBSS and transfer them to a glass plate. Using cataract knives, dissecton needles or Swann-Morton blades trim them and cut them to an appropriate size, now exceeding 2 × 2 × 2 mm. If possible, place the glass plate on a piece of dark paper. This will provide a good background for the tissues and facilitate dissection. It is essential that during dissection the tissue is kept moist with HBSS.

(vi) Transfer one to three explants with cataract knives or a wide mouthed pipette to the clot. In the latter case, the explants are sucked up with HBSS, deposited on the clot and the excess fluid is sucked off with the extra fine pipette.

(vii) Transfer the Petri dishes containing one or two watchglasses to an incubator kept at 37.5°C. Do not place Petri dishes directly on the metal shelves but interpose a piece of wood. This will prevent condensation of water on the inner side of the Petri dish lid which may drop on the explants and damage them.

3.3.3 *Renewal of Medium*

The medium has to be renewed every 2 − 3 days for avian tissues and every 3 − 4 days for mammalian tissues. For this purpose, fresh clots are prepared as described above.

(i) Gently detach the explants from the old plasma clot and lift off with the aid of two cataract knives.

(ii) Transfer them to, and wash them in, HBSS containing 2% serum contained in the cavity of a cavity slide.

(iii) After washing, suck the explants up with a wide-mouthed pipette and deposit them on the fresh plasma clot.

(iv) Separate and spread them out on the clot using a dissecting needle and cataract knife.

(v) Suck off the excess fluid with the extra fine pipette.

3.3.4 *Combination of Clot and Raft Techniques*

In this modification, the explants are first placed on strips of lens paper (see Section 7.1) or rayon acetate moistened with HBSS and these strips deposited on the clot. This method does not impair the viability or growth of the explanted tissues. It prevents the tissue from sinking into the pool of liquefied plasma and simplifies the medium changes. Instead of lifting off the explants individually, the rafts carrying the explants can be lifted *in toto*, washed with HBSS, drained on pieces of sterile filter paper and then placed on the fresh clot.

The watchglass technique has been eminently suitable for developmental studies on avian bone rudiments (6,26,27), for studies on differentiation of avian epidermis and oesophageal epithelium and its modulation by vitamin A (28,29). Further, it has proved very useful for investigating the action of carcinogenic agents, steroid hormones and vitamin A and their interaction in embryonic and adult, human, murine and rat tissues. Using the technique it could also be demonstrated that steroid hormones were necessary for the maintenance of their target organs and that carcinogenic agents induced pre-cancerous changes which could be inhibited by vitamin A (30−32).

The technique has certain limitations. The explants often liquefy the clot and sink into a pool of liquefied medium which may impair their oxygen supply. Consequently, the medium has to be changed frequently. The complexity of the medium rules out biochemical analysis and the duration of culture is somewhat limited. Although in many experiments the answer to a particular problem may be obtained within a short time, others may require a longer period of cultivation. Using the plasma clot technique the period of cultivation has, usually, not exceeded 4 weeks (33).

4. THE MAXIMOW SINGLE SLIDE TECHNIQUE

This is a modification of Fell and Robison's watchglass technique (6). It is derived from the double coverslip technique by which cells were grown on a plasma-extract clot (34) and it has been successfully adapted by Hardy (35) for developmental studies in foetal organs or organ rudiments (*Figure 4*). Hardy recommends the method for workers on a limited budget and with limited facilities and when the main effects to be examined are changes in morphogenesis or ciliary and secretory activity. Any changes can be observed by conventional light microscopy or by polarised or u.v. light microscopy.

4.1 Materials

In addition to those described in Section 3 the following are required.
Microconcavity slides 75 × 45 mm, 6−7 mm deep
Square coverslips, No. 3, 40 × 48 mm
Coverslip forceps
Safety razor blades, and a scalpel
Vaseline and Paraplast, four parts of Paraplast mixed with one part of Vaseline for sealing coverslips to the chambers
Fine natural bristle paint brush, for applying the Paraplast-Vaseline mixture.

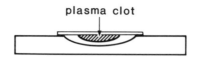

plasma clot

Figure 4. Maximow single slide technique, Hardy, 1978 (35).

4.2 **Setting up of Cultures**

(i)   Using fine knives and dissecting needle, prepare organs or organ rudiments to pieces of appropriate size (not exceeding 2 mm × 1.5 mm × 1 mm) in HBSS.

(ii)   With coverslip forceps place 40 × 48 mm coverslips into Petri dishes.

(iii) With a Pasteur pipette place three drops of chicken plasma into the centre of the coverslips. Stir the plasma with a glass rod to cover a circle of 25 mm.

(iv)   With a second pipette add one drop of chick embryo extract (50%) to the plasma, mix the extract and plasma thoroughly and quickly with the stirring rod.

(v)   When the mixture has clotted, place one to three explants on the clot, usually with the stromal surface attached to the clot.

(vi)   Transfer the explants with a scalpel blade or with the use of a wide-mouthed pipette. In the latter case, suck off the explant and HBSS and deposit on the clot. Remove the excess HBSS with the extra fine pipette.

(vii) Invert the coverslips carrying the clot and explants with a coverslip forceps over a Maximow slide. Seal the chambers with three coats of the Paraplast-Vaseline mixture, heated to 60°C.

(viii) Incubate the slide chambers at the desired temperature.

4.3 **Renewal of Medium**

(i)   Prepare fresh clots according to the method described above. Remove the paraffin seal from the slide chamber with a warmed scalpel or razor blade.

(ii)   Invert the coverslip on the stage of a dissecting microscope. Using fine forceps and a dissecting needle remove explants from the clots and transfer them to the slide well for rinsing with HBSS (two changes, 5 min).

(iii) Transfer each explant to fresh clot. Invert the coverslips over the Maximow slides and seal chambers as before.

Hardy (35) found the method suitable for studies on hair growth and differentiation of foetal mouse skin (36), for the study of hormonal effects on newly born mouse vagina (37) and the differentiation of foetal mouse gonads (38).

The method was less suitable for highly motile tissues such a foetal intestine and highly secretory organs. Although the method allows examination by light microscopy of the living explants the optics are not perfect and the tissues are usually too thick for examination by phase contrast.

5. AGAR GEL TECHNIQUES

These techniques eliminate some of the disadvantages of the plasma clot, avoiding clot liquefaction and allowing the use of defined media. The original technique was introduced by Spratt (10) who used a 3% Ringer solution albumen gel and later a 4% gel consisting of bicarbonate buffered saline solution, glucose, 11 amino acids, 10 vitamins and agar (11). Woff and Haffen (12) modified Gaillard's technique (9) to study development and differentiation of amphibian, avian and mammalian organs or organ rudiments. They used an agar gel contained in an embryological watchglass. Initially, their medium consisted of 1% agar, dissolved in Gey's solution mixed with 50% chick embryo extract and Tyrode solution. In later versions, chicken or horse serum was added to the

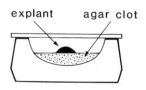

explant   agar clot

Figure 5. Embryological watchglass with agar clot, Wolff and Haffen, 1952 (12).

agar, or the embryo extract was replaced by certain amino acids and vitamins (39).

Wolff (40) found that foetal organ rudiments often became encapsulated by cells migrating from the cut surface of the explants. To prevent this he wrapped them in a piece of vitelline membrane obtained from hens' eggs. '

The method has also been applied to studies of tumour growth. Wolff *et al.* (41) succeeded in growing cells from human tumour cell lines or fragments of fresh human tumours on agar clots associated with chick mesonephros or on clots reinforced with chicken liver dialysate.

The method has been and still is widely used, although the composition of the agar gel has undergone some changes and some workers have replaced the saline solutions used to dissolve the agar with defined media.

In Section 5.1 a modification of the technique using such defined medium is described.

5.1 **Embryological Watchglass Technique** (*Figure 5*)

5.1.1 *Materials*

Embryological watchglasses, outer diameter 3.9 cm, inner diameter 3.5 cm
Glass lid 3.9 cm in diameter to cover watchglass
Paraffin wax heated to 60°C to seal lid on watchglass
Small natural bristle paint brush to seal lid with paraffin
Gibco agar 2%
Double strength Morgan, Morton and Parker's medium 199 based on Hanks' salts
Single strength medium 199, based on Hanks' salts
Newborn or foetal bovine serum, inactivated at 56°C for 30 min
HBSS
Pasteur pipettes
Wide-mouthed pipettes
Very fine Pasteur pipettes made in the laboratory by drawing out Pasteur pipettes
  to a fine point in flame
Pipette rack
Cataract knives
Watchmaker's forceps, fine curved scissors, dissecting needles
Glass plates, approximately 8 cm square, for dissecting organs.

5.1.2 *Preparations of Agar Gel*

(i)   Heat 2% agar in a water bath kept at 90°C until liquefied, then mix quickly in a glass tube with equal amounts of double strength medium 199, kept at 37°C.

(ii) When thoroughly mixed, place 0.5 ml in the bottom of the watchglasses, add

0.1 ml foetal bovine or newborn calf serum and 0.4 ml single strength medium 199, again mix thoroughly and quickly to avoid premature clotting. The mixture begins to clot at 36°C and is firmly set at 30°C.

If a smaller clot is desired the proportions of agar, serum and medium 199 can be modified accordingly. In experiments which involve a study of hormonal action or that of other agents, the compounds to be investigated are incorporated into the clot by adding them to the single strength medium 199.

5.1.3 *Setting up of Cultures*

(i)   Remove the organs or organ rudiments from the embryos with the aid of a dissecting microscope. During dissection, keep the embryos moist with HBSS containing 2% serum.
(ii) Transfer the tissues to be cultured and wash them in HBSS contained in a cavity slide.
(iii) Transfer them to the agar gel, either with the aid of two cataract knives or by gently sucking them up in a wide-mouthed pipette with HBSS and deposit them on the agar.
(iv) Orientate the explants on the agar with two needles and suck off the excess fluid with a fine pipette. Usually, one explant is accommodated on each gel. The optimum size of the explant is 1.5 x 1.5 x 10 mm, the maximum size 2 x 2 x 1.5 mm. However, if the tissue is very thin, say less than 0.5 mm thick, the explant could be larger.
(v)   Using the small paint brush, seal the glass lids onto the chambers with at least two coats of warm paraffin wax (60°C). Transfer the chambers to the incubator and incubate at 37°C or any other desired temperature. It is advisable not to place them directly on the metal shelves but to interpose a piece of wood.

Many experiments involving studies of morphogenetic changes and cytodifferentiation do not require a lengthy culture period and the answer to the question posed may be available within 3 – 7 days. In this case, a change of medium is not necessary. If the experiment requires a longer period of cultivation, the explants have to be moved to a fresh agar gel every 5 – 7 days. A fresh gel is made in a fresh chamber, prepared with the same proportion of agar, serum and defined medium as before.

Remove the lids from the old embryological watchglasses with a warmed razor or scalpel. Lift the explant off the old clot with the aid of two cataract knives or one cataract knife and a dissecting needle, wash in HBSS contained in a cavity slide, transfer to the fresh clot and seal the watchglasses with a fresh sterile glass lid, using the small paint brush and paraffin wax.

The viability of the explants and macroscopic changes can be monitored during cultivation by viewing them by daylight or with the aid of a light source from the dissecting binocular. Healthy and growing tissues, usually, appear translucent with a shiny surface while opacity suggests loss of viability or the beginning of necrosis of the explanted tissue.

This modification in which the saline solution was replaced by defined medium has been found suitable to study the induction by testosterone of the prostate gland from their 'anlage', the urogenital sinus, from mouse or rat embryos and the role of mesenchyme in this process (42,43). The agar supported the development of the prostate gland from the whole sinus as well as that of its components, epithelium and mesenchyme, separated and re-associated with each other in various recombinations.

6. EMBRYO CULTURE FOR TERATOLOGICAL STUDIES

New (44,45) and Cockcroft (46) have developed a method by which whole mouse and rat embryos can be grown in culture. As very early pre-implantation embryos have a great capacity for recovery (44) they are less suitable for teratological studies than older embryos which are more susceptible to the influence of toxic agents.

New and Cockcroft chose, therefore, embryos of two later stages of gestation: 9.5 days and 11.5 days. The 9.5 day embryos consist of a pre-somite cylinder, 1.5 mm long and 0.6 mm in diameter. The 11.5 day embryo has $27-31$ somites and is $3-3.8$ mm in length.

6.1 **Explantation Procedure**

Remove the pregnant uteri and open in saline solution along the anti-mesometrial side and cut away the conceptus with a cataract knife. After transfer to fresh saline, remove the maternal decidua and open Reichert's membrane. The ectoplacental cone or, in older embryos, the allantois is left attached to the embryos which are now ready to be placed into their culture chambers.

Two types of culture chambers have been used.

(i) *Cylindrical 60 ml bottles*. These are partially filled with medium and partially with a gas mixture. The bottles are laid horizontally on rollers and rotated at $30-60$ r.p.m. during incubation which ensures maximum and even exposure to the agent under investigation.

(ii) *Rotator*. Small culture bottles are attached to a hollow rotating drum. The gas mixture is passed continuously through the drum. This provides a constant level of CO_2 and oxygen with a minimum variation of pH of the medium.

6.2 **Gas Mixture**

For embryos explanted at 9.5 days' gestation the gas mixture consists of 5% CO_2, 5% oxygen and 90% nitrogen for the first 24 h. After this period a mixture of 5% CO_2, 20% oxygen and 75% nitrogen is used.

Embryos explanted at 11.5 days' gestation require 5% CO_2 and 95% oxygen throughout the culture period.

6.3 **Medium**

The medium consists of rat serum. This is obtained from ether-anaesthetised donor rats. A better result is obtained if the blood is centrifuged immediately and before it has formed a clot. After centrifugation, the clot formed in the supernatant is gently squeezed to release the serum underneath. This is re-centrifuged, decanted and stored at $-10°C$. Before use, the serum is thawed, heat inactivated at 56°C for 30 min, gassed to remove any residual ether and centrifuged to remove any further fibrin clots.

6.4 **Protocol**

Six 9.5 day embryos are placed in each 60 ml bottle with 6 ml serum. Three 11.5 day embryos are explanted per bottle in 9 ml serum. At both ages, they are incubated at $37 - 38°C$. The embryos are explanted either floating freely in the serum or placed on collagen-coated fabric attached to a coverslip.

A high proportion of the embryos develop well in culture. The younger embryos form $25 - 28$ somites and their protein content rises significantly, close to the pattern of development seen *in vivo*. Embryos explanted at 11.5 days' gestation develop at a slightly slower rate than *in vivo* but they also show increased somite number and protein content. Embryos anchored in position can be inspected during cultivation and their development followed. At the same time, the influence of toxic agents on normal development can be monitored. In this way, the effect of CO_2, excess glucose and hyperthermia has been successfully studied (46). The method is equally suitable for investigating the influence of other toxic or suspect agents on normal development.

7. GRID TECHNIQUE

This method combines the use of fluid defined or semi-defined media with a firm support of the explanted tissues which prevents them being submerged into the medium and thus becoming deprived of oxygen. The importance of oxygen for the growth of adult tissues was recognised by Trowell who designed an apparatus in which cultures explanted on a grid were enclosed in a chamber under an atmosphere of carbon dioxide and oxygen $(17 - 19)$. As currently used, the metal grid for supporting cultures in fluid medium is basically unchanged but the original gassing chamber introduced by Trowell has been considerably modified and simplified.

This modified Trowell technique is now widely used for embryonic and adult tissues and is described here (*Figures 6* and *7*).

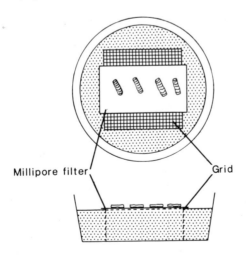

Millipore filter   Grid

Figure 6. Modified Trowell technique.

Figure 7. Modified Trowell technique: Petri dish with culture chambers, grids and Millipore filter strips.

7.1 **Materials**

Sterile distilled water or HBSS

Petri dishes 91 mm in diameter

Round sterile filter papers to fit into the bottom of the Petri dishes with two 3.5 cm holes punched in them

Plastic culture chambers 32 mm in diameter (Sterilin)

Lens paper (Gurr, London), degreased in ether and alcohol before use (see Section 7.1.1 for method of preparing lens paper)

Millipore filter 0.5 μm pore size (Millipore Corp., Bedford, MA, USA) both cut into strips measuring 1 × 2.2 cm before sterilisation. *Note* It may be worthwhile trying polycarbonate membranes (Nucleopore, Sterilin) which are transparent and detergent free instead of Millipore filters. Explants grown on such membranes could be inspected during cultivation. They may be particularly suitable for the cultivation of small whole organs such as foetal mammary glands which could be fixed, stained and examined on the membrane thus eliminating sectioning.

Grids of expanded metal (Expanded Metal Co., Hartlepool, UK) 18 mm square, 4 mm high, cut out from a larger sheet with a pair of scissors. The ends are bent down to form a bridge or table. After using a metal grids immerse them in concentrated nitric acid and leave for 24 − 48 h. Remove grids from acid and transfer them to running tap water for at least 24 h. Rinse in three changes of distilled water and dry in drying oven. The nitric acid can be used several times.

Medium 199 based on Earle's salts

Medium 199 based on Hanks' salts (holding medium)
Insulin (Sigma)
Newborn calf serum, heat-inactivated
Universal containers, glass or plastic, 20 ml capacity
Bijou bottles, glass, 5 ml capacity
Graduated pipettes
Pasteur pipettes
Pipette rack
Penicillin (Glaxo)
Streptomycin (Glaxo)
Fungizone (Flow Laboratires, Irvine, UK)
2 Straight forceps approximately 10 cm long
2 Straight scissors approximately 10 cm long
2 Straight fine scissors
2 Watchmaker's forceps No. 3 or 4 (Weiss, London, UK)
2 Cataract knives
2 Swann-Morton blades No. 10 − degreased with absolute ethanol before use
Glass cavity slides
Glass plate, approximately 8 cm square, for dissection
70% Alcohol
Stainless steel rack
Anaerobic Macintosh and Fildes Jar (modified)
Gas cylinder containing 5% CO_2 and 95% oxygen.

7.1.1 *Preparation of Lens Paper for Culture*

Lens paper is normally covered with a fine layer of grease which has to be removed before use.

(i)   Cut lens paper into pieces approximately 5 x 8 cm in size and put them in a Petri dish.

(ii)   Immerse in ether for 1 h, change ether and leave for a further hour.

(iii)   Take ether off and immerse into absolute ethanol for 1 h, change absolute ethanol and leave for a further hour. Take absolute ethanol off and immerse into glass-distilled water, changing the water about 8 − 10 times. Leave lens paper in glass-distilled water overnight. Dry in drying oven (37°C).

(iv)   Before use, cut into strips of required size and sterilise with dry heat or by autoclave.

7.2 **Explantation Procedures**

(i)   Assemble the apparatus necessary for cultivation and prepare the media before dissecting the organs.

(ii)   Moisten three layers of filter paper inserted into each Petri dish before sterilisation with 10 ml of sterile distilled water or HBSS.

(iii)   Place two plastic culture chambers into each dish over the holes punched into the paper, and place the expanded metal grids into the chambers.

(iv)   Fill another Petri dish with holding medium 199 (based on Hanks' salts) containing 2% newborn calf serum for washing and keeping the tissues moist before and during dissection.

The exact procedures vary with different tissues and those used for the explantation of prostate gland, trachea and skin are described in some detail.

7.2.1 *Prostate Gland*

This is derived from mice or rats, 6 weeks to 6 months old.

(i)   Secure the animals with map pins to a cork mat and thoroughly wash their fur with 70% alcohol.
(ii)   Open the abdominal skin with a pair of straight scissors with one horizontal and two vertical cuts. Secure the resulting skin flap with a map pin leaving the area around the bladder exposed.
(iii)   Lift up the bladder with a pair of watchmaker's forceps and cut its adhesion to the underlying muscle. Turn it over to reveal the prostate gland underneath. The gland is distinguishable from the surrounding tissues by its shiny surface and mother of pearl colour.
(iv)   Lift the gland up with a pair of watchmaker's forceps and cut from the bladder base with a pair of fine scissors.
(v)   Rinse the gland with ice cold holding medium, transfer to a cavity slide filled with the same medium, and keep on ice until required for dissection. At least six glands are required for one group of experiments.
(vi)   Transfer the glands to a glass slide for dissection.
(vii)   First free them from the surrounding fat and connective tissue with a pair of Swann-Morton knives.
(viii)   Separate the two lobes of the gland by cutting through the connective tissue between them.
(ix)   Then subdivide them, not by cutting them as this would result in regenerative hyperplasia at the cut edges, but by teasing them gently apart into smaller lobules. The small lobules measuring $2 \times 2 \times 2$ mm are now ready for explantation.

Care should be taken to keep the tissue moist during dissection, otherwise it will dry out and stick to the glass.

7.2.2 *Trachea*

This is obtained from neonatal mice or rats, $2-5$ days old.

(i)   Secure the animals on a small cork mat with pins and wash thoroughly with 70% alcohol.
(ii)   Open the skin and rib cage with a pair of fine scissors and expose the trachea by pulling the ribs apart. The trachea is easily recognised by its cartilage rings.
(iii)   Loosen it with the aid of two watchmaker's forceps. One holds the organ while the other slides underneath it and separates it from the underlying oesophagus by moving the forceps to and fro.
(iv)   When this is accomplished, lift it up and remove from the rat with a pair of fine scissors and watchmaker's forceps.

(v)   Rinse the organ with ice-cold holding medium, deposit in a cavity slide filled with the same medium and keep on ice until further dissection.

(vi)   Before explantation, transfer the tracheas to a glass slide, trim at the proximal and distal ends and halve with the aid of two Swann-Morton knives. The resulting halves measure approximately 1.5 – 2 mm in length and are explanted as tubes.

7.2.3 *Skin*

This is usually obtained from 17 – 18 day mouse or rat embryos.

(i)   Secure the embryos on a small cork mat with pins and moisten their skin with cold holding medium.

(ii)   With the aid of a pair of watchmaker's forceps and a pair of fine scissors remove the dorsal skin *in toto*, rinse with cold holding medium and transfer to a glass slide. Once removed from the embryo the skin contracts, tends to curl up and has to be straightened repeatedly during dissection.

(iii)   Scrape the underside free of fat, if present, so that the skin now consists of two main layers: dermis and epidermis.

(iv)   Subdivide it with the aid of two Swann-Morton knives into fragments measuring 3 x 3 x 4 mm.

7.3 Medium

Several media have been examined as to their efficacy in maintaining structure and function of the explanted organs. These include Ham's F12 medium, Waymouth's medium 752/1, medium CMRL 1066, medium 199 and Trowell's T8 medium.

Trowell's medium seems suitable only for relatively short-term maintenance of adult tissue. Besides, it contains a high concentration of insulin (50 μg/ml) which may interfere with the action of hormones and mask the effects of other growth factors to be studied. It is, therefore, not suitable.

As no basic difference has been observed between other media, medium 199 is used routinely, either alone or supplemented with serum and small amounts of insulin.

The composition of the medium used is as follows:

Medium 199 based on Earle's salts

Heat-inactivated newborn calf serum, 5% for hormonal studies, 10 – 15% for studies on carcinogenesis

Insulin 0.5 – 1.0 μg/ml

Antibiotics and anti-fungal agents: penicillin 250 I.U./ml, streptomycin 100 μg/ml, fungizone 1.4 μg/ml.

7.4 Explantation

(i)   Place the strips of lens paper or Millipore filter on a glass slide or the bottom of a Petri dish and moisten with 1 – 2 drops of holding medium.

(ii)   With the aid of two cataract knives, transfer 6 – 8 fragments of prostate gland or four halved tracheas to the strips; lens paper for adult prostate or Millipore filter for neonatal trachea. If lens paper is not available Millipore filter can be used for the prostate glands as well.

(iii) Transfer the strips holding the explants to the grids in the culture chambers with a pair of watchmaker's forceps and fill the chambers with approximately 3 ml of medium which should reach the upper level of the grids and keep the surface of the explants moist without submerging them.

The explantation of skin is a little difficult as it tends to curl up. This can be overcome by pressing a moistened strip of Millipore filter onto the dermal side of the skin explants. This makes them flatten out and adhere tightly to the substrate. The strips are inverted before placing them on the grids so that the skin is growing with the epidermis uppermost.

Compounds to be investigated can be added to the medium, at the desired concentration, at this stage. Alternatively, they can be incorporated into the medium first and the mixture kept at −20°C until used. This latter method ensures a tighter control of experimental conditions and is more labour saving.

7.5 Incubation and Perfusion

(i) Stack the Petri dishes holding two culture chambers each in a rack of stainless steel and place into a Macintosh and Fildes Pattern Anaerobic Jar (Baird and

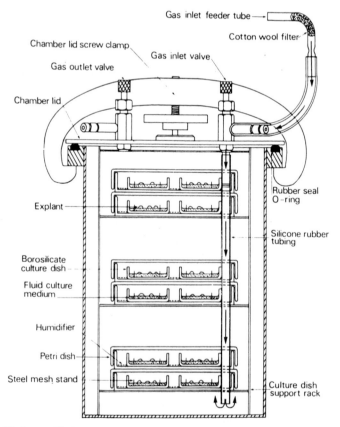

Figure 8. Modified anaerobic Macintosh and Fildes jar for gassing cultures.

Tatlock Ltd., London, UK) (*Figure 8*) modified by Fainstat (47). The attachments meant for anaerobic control are discarded and the lid fitted with two valves.

(ii)   Place the jar in the incubator and connect to the gas cylinder with silicone rubber tubing running through a small opening at the top of the incubator.

(iii)   Perfuse the jar with a mixture of 5% CO_2 and 95% oxygen *via* a sterile filter. For cultures of adult prostate perfuse for 25 min at a flow-rate of 150 ml/min; for neonatal trachea and skin, for 20 min at a flow-rate of 125 ml/min. Check the flow-rate and quantity of the gas mixture with the use of a flow meter.

(iv)   Seal the jar tightly so that the concentration of the gas mixture inside the jar remains constant until it is opened for medium renewal.

(v)   Incubate the cultures at 37.5°C.

The mixture of 5% CO_2 and 95% oxygen has been found suitable for the cultivation of prostate glands, trachea and skin and prevents the central necrosis of the tissue which may occur if cultured in air. However, in other tissues, a lower proportion of oxygen may be as effective as long as its concentration is above that in air.

7.6 Renewal of Medium

The medium is renewed every 2 − 3 days.

(i)   Remove the Petri dishes holding the cultures from the Macintosh jar.

(ii)   Suck off the old medium with a Pasteur pipette and replace it with the same quantity of fresh medium. The culture chambers can remain in the same Petri dish if the period of culture is relatively short. For prolonged cultivation, it is recommended to transfer the chambers to a new set of sterile Petri dishes, moistened with sterile water or HBSS.

(iii)   Following the change of medium, return the Petri dishes holding the culture chambers to the Macintosh and Fildes jar.

(iv)   It should again be tightly sealed and re-perfused with the same gas mixture as before.

The grid technique has proved successful in studying the growth and differentiation of embryonic and adult tissues, in investigating the effect of steroid hormones on their target organs and in the interaction of carcinogens and smoke condensate with vitamin A.

Studies using rat prostate glands have shown that the organs remain androgen dependent and responsive *in vitro* and that they metabolise testosterone to the active androgen, dihydrotestosterone, as measured by steroid chromatography of tissue and medium (48). Human benign prostatic hyperplasia could be well maintained in organ culture (49) and testosterone was found to increase RNA synthesis of its epithelium and stroma (50). Using the grid technique, rat and human bladder have been maintained for prolonged periods and the urothelium was seen to differentiate in a manner similar to that *in vivo* (51,52).

In mouse or rat prostate glands, human and murine embryonic lung, and embryonic and neonatal rat trachea, carcinogenic hydrocarbons and cigarette smoke condensate induced, within a short time of exposure, changes of a precancerous nature. These could be inhibited or reversed by vitamin A and its structural analogues (53 − 56).

8. LONG-TERM CULTIVATION OF HUMAN ADULT TISSUES

With many organ culture methods, growth and differentiation of the explanted organs can usually be maintained for only limited periods *in vitro*. In many experiments the answer to a particular problem can, in fact, be obtained within days or weeks of cultivation, but in others a longer term cultivation may be necessary. This is particularly true for studies of carcinogenesis in human tissues which require a prolonged exposure to carcinogenic agents.

With this aim in mind, a method has been introduced by which human tissues were intermittently exposed to fluid medium and an appropriate gas phase. Using this technique, the characteristic growth pattern and differentiation could be maintained in culture for prolonged periods.

The tissues include bronchus (20), breast (21), oesophagus (22) and uterine endocervix (23).

8.1 Explantation and Cultivation

(i)   Specimens can be obtained from immediate autopsies or surgical resections. They should be dissected from the surrounding tissues and transported to the laboratory, usually immersed in ice-cold Leibowitz medium. However, other media are equally suitable as long as they are well buffered. To prevent bacterial infection of the tissues it is recommended to add antibiotics to this holding medium, and after arrival in the laboratory, transfer the tissue to a similar medium for further washes. It is recommended to use a concentration of antibiotics at least twice that used for cultivation (e.g., 200 I.U./ml penicillin, 200 μg/ml streptomycin).

(ii)   Trim the tissues further, free from fat, stroma and underlying muscle and divide into fragments of a size suitable for explantation. The measurements used vary for different tissues. For instance, bronchial explants measure 2 cm^2, breast and endocervix $0.5 - 1.0$ cm^2 and oesophagus 1 cm^2.

(iii) Place the explants on the bottom of 60 mm tissue culture grade plastic dishes to which they adhere and then submerge them in $3 - 5$ ml medium. Place organs composed of epithelium and underlying stroma into the dish with the stromal side attached to the bottom of the dish with the epithelium at the free surface.

The number of explants per dish varies with different tissues and authors. Thus $6 - 9$ explants of endocervix, $2 - 6$ explants of breast tissue and $6 - 18$ explants of oesophagus can be grown in each dish. However, it may not be necessary to follow this regimen exactly and other workers may find a different explant size and number per dish more suitable for their particular requirements.

8.2 Medium

The medium found to be most suitable for the long-term maintenance of human tissues in organ culture was medium CMRL 1066, supplemented routinely with foetal bovine serum, hormones and antibiotics. The composition of the medium is as follows.

Medium CMRL 1066 (Gibco)
5% Heat-inactivated foetal bovine serum
$2 - 4$ mM L-glutamine (Flow Laboratories)
1 μg/ml Bovine insulin (see *Table 1*) (Sigma)

0.1 µg/ml Hydrocortisone hemisuccinate (see *Table 1*) (Sigma)
Penicillin 100 – 300 I.U./ml (Glaxo)
Streptomycin 100 – 300 µg/ml (Glaxo)
Fungizone (Squibb) 5 µg/ml or Gentamicin (Schering) 100 µg/ml.

For the cultivation of explants of hormone-dependent organs specific hormones should be added to the medium [e.g., for breast tissue, progesterone, oestradiol, aldosterone and prolactin have been applied, either continuously or as a 28 day regimen to mimic the menstrual cycle (21)].

8.3 Incubation

The explants are incubated at a temperature of 36.5°C.

(i)   Place the dishes containing the explants in 3.5 ml medium on individual trays which can hold up to 60 dishes, and transfer these to an atmosphere controlled chamber (Bellco).

(ii)   Perfuse the chamber with a gas mixture of 5% CO_2, 45% oxygen and 50% nitrogen.

(iii)   Then place on a rocker platform and rock at 3 – 10 cycles per min, depending on the tissues used. This causes the medium to flow intermittently over the surface of the explants.

The assembled unit will fit into most standard incubators.

8.4 Change of Medium

The medium is changed two or three times weekly, the gas mixture is replaced after each medium change and the atmosphere controlled chamber returned to the rocker platform and the incubator.

Using this technique the tissues could be maintained in culture for prolonged periods with only minor alterations of their normal growth pattern (*Table 1*).

Table 1. Long-term Culture of Human Explants.

	Length of Cultivation time up to (months)	Results
Bronchus (20)	4	After 6 weeks, reduction in height of columnar epithelium, some loss of goblet cells, some metaplasia.
Breast (21)	6	Lobular architecture and typical ductal epithelium preserved. Secretory activity present with changes related to menstrual cycle.
Oesophagus (22)	4	Characteristics of squamous epithelium retained as demonstrated by light and electron microscopy.
Endocervix (23)	5	Epithelium viable as seen by mitotic activity and ultrastructure. After 4 weeks partial loss of mucus secretion with focal metaplasia.

Insulin is a growth factor and in experiments where changes in cell proliferation are a criterion of effect the presence of insulin may mask the effect. It is recommended to omit it for such experiments or reduce the dose (e.g., to 10 ng/ml).
Hydrocortisone at the concentration recommended (0.1 µg/ml) inhibits the action of carcinogenic agents (57) and is not suitable for studies of carcinogenesis. It should be omitted or used at a 20-fold lower concentration.

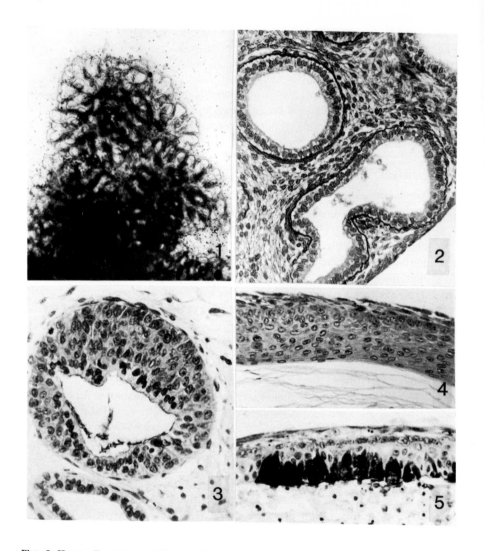

Plate I. Human Foetal Lung. Effects of 3,4 Benzpyrene.

Figure 1. Living culture of human foetal lung after 2 weeks growth in control medium, showing organised growth. Note spreading of new bronchioli at edge of explant, × 19.

Figure 2. Bronchioli at higher magnification after 2 weeks growth in control medium. Azan substitute, × 243.

Figure 3. Induction of hyperplasia. Hyperplastic bronchiolus seen after 2 weeks growth in presence of 3,4-benzpyrene, showing small crowded cells and superficial layer of secretory epithelium. Haematoxylin-eoxin, × 342.

Embryonic Rat Oesophagus, Mucus Transformation of Epithelium by Vitamin A.

Figure 4. Oesophagus grown for 1 week in control medium. Note thickness of epithelium and keratin fibres. PAS with diastase digestion, × 281.

Figure 5. Oesophagus grown for 1 week with excess vitamin A. Note absence of keratin and formation of mucus-containing goblet cells. Mucicarmine, × 281.

9. PREPARATION OF ORGAN CULTURES FOR LIGHT MICROSCOPY

The basic procedures to prepare histological sections for light microscopy are the same as those used for other tissues but the timings are shorter.

9.1 Materials

Fixative: Bouin's or 3% acetic Zenker
Ethanol series 50 to 100%
1% Eosin in 70% ethanol
Cedar wood oil
Soft wax, melting point 45 – 56°C (Raymond Lamb, Sunbeam Road, London)
Hard wax, melting point 63°C (Raymond Lamb, Sunbeam Road, London)
2 Needles
2 Cataract knives
Glass watchglasses 5 cm in diameter
Wide-mouthed pipettes
Hotplate.

9.2 Fixation Dehydration and Embedding

(i)   At the end of the culture period, lift the strips carrying explants from medium. Rinse them with cold HBSS.
(ii)   Drain on filter paper.
(iii)   Immerse the strips carrying cultures in fixative for 35 – 40 min.
(iv)   Following fixation in Bouin, transfer the strips to several changes of 70% ethanol.
(v)   Following fixation in Zenker, transfer to tap water for 1 h, 50% ethanol for 1 h, 70% ethanol for 1 h, 70% ethanol overnight, 90% ethanol for 1 – 1.5 h.
(vi)   Remove explants from the strips with cataract knives.
(vii)   Add a few drops of 1% eosin to 95% ethanol. Transfer the explants to 95% ethanol with eosin for 30 min.
(viii) Two changes of absolute ethanol, 1 h each. Cedar wood oil at least 1 h; cedar wood oil overnight; soft wax 1 – 1.5 h; soft wax 1.5 – 2 h. Two changes of hard wax, 30 min – 1 h each.

9.2.1 *For Embedding*

(i)   Warm the watchglass on the hotplate.
(ii)   Partially fill with hard wax.
(iii)   Transfer the explants and wax to the watchglass with a wide-mouthed pipette.
(iv)   Orientate the explants in wax with two needles before it solidifies. When solidified, plunge the watchglass with embedded explants into cold tap water. This helps the wax block to detach from the watchglass.
(v)   Block up and cut serial sections at 5 – 6 μm.
(vi)   Store the sections for at least 1 day before staining.

9.3 Staining Methods

Before staining, the sections have to be de-waxed in xylene and taken down to distilled water *via* immersion in alcohols of descending concentrations. *N.B.* As xylene vapour is toxic and may be carcinogenic it is best handled in a fume cupboard.

(i)   Xylene 5 min each.

(ii)   Xylene.

(iii)   At this point make sure that the wax is completely removed. If not, return to xylene until it is gone, and transfer to: absolute ethanol; 90% ethanol 5 min each; 70% ethanol; 50% ethanol; distilled water.

9.3.1 *Haematoxylin-Eosin (H-E)*

This is a routine stain for histology and histopathology and much of the present knowledge of normal and morbid histology has been gained by the study of H-E stained sections.

(i)   *Protocol*

(a)   Ehrlich's haemotoxylin 5 – 15 min.

(b)   Distilled water, quick dip.

(c)   Remove excess stain by differentiating in acid alcohol (1% hydrochloric acid in 70% ethanol) for a few seconds. The blue stain of haematoxylin is changed to red by the action of the acid.

(d)   Distilled water, quick dip.

(e)   Regain blue colour and stop decolorisation by washing in alkaline running tap water for 5 – 15 min at room temperature, stain in 1% aqueous eosin for 1 – 3 min.

(f)   Wash off surplus stain in distilled water, dehydrate in alcohols and clear in xylene bearing in mind that eosin is removed from the tissues by water and low grade alcohols and less readily by absolute alcohol. The degree of staining is thus easily controllable and any overstaining can be remedied during the passage through the alcohols. The recommended times are: 2 – 3 sec each for 70% ethanol and 90% ethanol; 30 sec for absolute alcohol; 1 – 2 min for xylene; 1 – 2 min or longer for xylene.

(g)   Mount in a synthetic resin such a DPX.

(ii)   *Results*

(a)   Nuclei, blue to blue-black.

(b)   Chromosomes, blue-black.

(c)   Nucleoli, purplish.

(d)   Cartilage, light to dark blue.

(e)   Basement membrane, pink.

(f)   Cytoplasm, various shades of pink.

(g)   Muscle fibres, thyroid colloid, elastic fibres, deep pink.

(h)   Collagen and osteoid tissue, light pink.

9.3.2 *Periodic Acid-Schiff (PAS)*

This method is used to demonstrate polysaccharides (glycogen), neutral mucopolysaccharides, mucoproteins, glycoproteins and glycolipids.

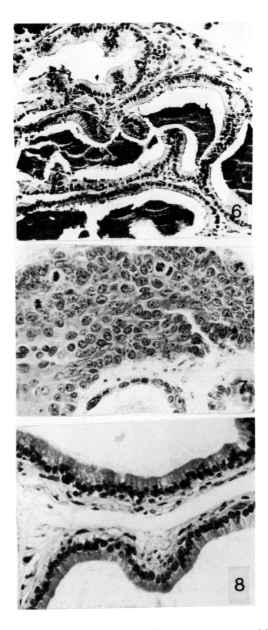

Plate II. Adult Mouse Prostate Gland. Effect of 20-Methylcholanthrene and its Reversal.

Figure 6. Mouse prostate grown for 3 weeks in control medium showing alveoli lined with one row of mucus-secreting epithelium. PAS with diastase digestion. × 243.

Figure 7. Hyperplastic alveolus in mouse prostate treated for 11 days with carcinogen, followed by 10 days growth in control medium. It shows small crowded cells and abundant cell division. Haematoxylin, × 304.

Figure 8. Alveolus in mouse prostate gland treated for 11 days with 20-methylcholanthrene followed by 10 days growth with a vitamin A analogue. Note reversal of hyperplasia. The lining epithelium consists of one row of columnar cells. Haematoxylin-eosin, × 410.

173

(i)   *Protocol*

(a)   After de-waxing bring sections to distilled water *via* alcohols in descending concentrations.

(b)   Oxidise for 15 min in 1% periodic acid.

(c)   Wash for 15 min in running tap water and rinse in distilled water.

(d)   Place in Schiff's reagent in the dark for 10—20 min.

(e)   Wash for 10 min in running tap water.

(f)   Stain nuclei for 10 min in Mayer's haematoyxlin. Do not use Ehrlich's haematoxylin which will also stain some PAS-positive tissue components.

(g)   Dehydrate for 2 min each in ascending concentrations of alcohol.

(h)   Clear in xylene 1—2 min each.

(i)   Mount in a synthetic resin, such as DPX.

(ii)   *Results*

(a)   PAS-positive substances red or magenta, nuclei blue.

For some experiments it is necessary to exclude glycogen. This can be removed by incubating the slides with diastase before staining them and the protocol is slightly modified accordingly.

(a)   Bring sections to distilled water.

(b)   Incubate them for 12 min with a 1% solution of malt diastase.

(c)   Rinse in distilled water.

(d)   Wash in running tap water for 5 min.

(e)   Transfer to 1% periodic acid for 15 min.

The remainder follows the protocol outlined above.

9.3.3 *Azan Substitute*

This stain provides a very clear differentiation between epithelium, stroma and cartilage with a well-defined basement membrane between epithelium and underlying stroma.

(i)   *Protocol*

(a)   De-wax sections and bring down to distilled water.

(b)   Place in carmalum for 20 min.

(c)   Wash in distilled water for 1 min.

(d)   Transfer to 1% phosphomolybdic acid for 5 min.

(e)   Wash in distilled water for 1 min.

(f)   Transfer to aniline blue for 10 min.

(g)   Wash in distilled water for 1 min.

(h)   Differentiate in 96% alcohol for 5 min.

(i)   Transfer to: absolute alcohol, 10 dips; absolute ethanol, 2 min; xylene, 5 min; xylene, 5 min.

(j)   Mount in a synthetic resin such as DPX.

(ii)   *Results*

(a)   Nuclei, red.

(b)   Chromosomes, red.

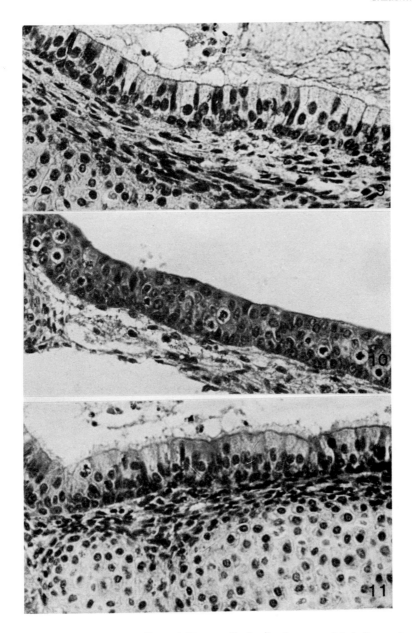

Plate III. Neonatal Rat Trachea. Effects of Cigarette Smoke Condensate and their Reversal.

Figure 9. Trachea grown for 18 days in control medium. The tracheal epithelium consists of columnar ciliated cells. Azan substitute, × 379.

Figure 10. Trachea grown for 14 days with smoke condensate followed by 4 days growth in control medium. The cilia are lost and the epithelium consists of large numbers of crowded small cells with abundant cell division. Azan substitute, × 379.

Figure 11. Trachea grown for 14 days with smoke condensate followed by 4 days growth in presence of a vitamin A analogue showing reversal of smoke condensate effects. Azan substitute, × 379.

175

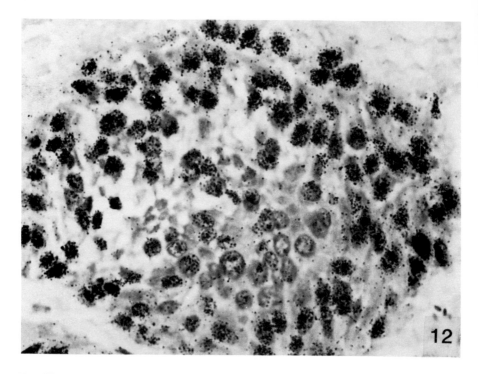

Plate IV.

Figure 12. Autoradiograph showing uptake of [³H]uridine by epithelium of human benign prostatic hyperplasia cultured for 7 days in the presence of testosterone, × 572.

(c)   Epithelial cytoplasm, pink.
(d)   Basement membrane, brilliant blue.
(e)   Collagen, blue.
(f)   Cartilage, blue.

10. AUTORADIOGRAPHY

Measurements of isotope incorporation of labelled tissues by extraction and counting give the total radioactivity but its distribution between different tissue components or cells is uncertain. Autoradiography provides this information and pinpoints the incorporation in different tissue components or individual cells, their cytoplasm or nucleus. This label can be recognised in suitably treated sections by light microscopy.

Autoradiographs were initially prepared by the method of Doniach and Pelc (58) using Kodak AR 10 stripping film but later a dipping method was used routinely. The latter method was faster and, because very thin films could be produced, resulted in better resolution. The following section describes the dipping method (for stripping film method see Chapter 9).

10.1 **Labelling**

The explants are removed from the strips of lens paper or Millipore filter and immersed in 2 ml of fresh medium warmed to 37°C.

(i) *DNA synthesis*: add [³H]thymidine 1 μCi/ml (37 KBq). Mix well and incubate cultures for 5 h.

(ii) *RNA synthesis*: add [³H]uridine 10 μCi/ml (370 KBq) to medium. Mix well and incubate for 40 min. With this short labelling time the uptake is confined mainly to rapidly labelling RNA in the nucleus. Longer labelling periods will show progressive cytoplasmic labelling from rRNA and transferred mRNA.

(iii) *Protein synthesis*: add [³H]leucine 10 μCi/ml (370 KBq). Mix well and incubate for 40 min.

10.2 Fixation

At the end of incubation, lift explants from medium and wash them thoroughly with HBSS to remove unincorporated radioactivity. For fixation, immerse them in acetic alcohol (1 part of glacial acetic acid, 3 parts of 80% alcohol) for 25 min. Transfer them directly to 10% formal saline, two changes, 1 h each or overnight.

10.3 Dehydration and Embedding

Follow the same procedure as that described in Section 9.2 except that tissue sections are cut at 4 μm.

10.4 Dipping Procedure

10.4.1 *Apparatus and Materials*

Ilford K5 nuclear research emulsion in gel form. Store in a refrigerator at 4°C. When in use it lasts for about 6 weeks; if unopened it lasts 3 months or longer
2 Small beakers
Glass or polypropylene cell and spoon
Forceps
Test slides
Pegs
Drying rack
Black plastic box
Silica gel
Glass-distilled water.

10.4.1 *Preparation of Emulsion*

(i)   Open emulsion in a darkroom illuminated with a Kodak red filter.
(ii)   Using a plastic or glass spoon, spoon enough emulsion for immediate requirements into a chemically clean beaker.
(iii)   Melt the emulsion in a water bath maintained at 40 – 42°C.
(iv)   When melted, fill 2/3 of glass cell with emulsion, then fill almost the to top with glass-distilled water warmed to 40 – 42°C.
(v)   Close the cell and invert it once or twice.
(vi)   Replace the cell in the water bath and allow it to sit for 1 – 2 min before dipping.

10.4.2 *Applying Emulsion*

(i)   Before dipping the slides, clear in xylene then transfer to 50:50 xylene ethanol, 100, 90, 70, 50, 30% ethanol and distilled water, 2 min each.

(ii)   Acid-soluble precursors of DNA, RNA or protein may be removed in ice-cold 10% trichloroacetic acid (2 x 10 min) followed by three rinses in ice-cold distilled water.

(iii)   Bring the slides in distilled water to the water bath, rinse once or twice with glass-distilled water, drain in an upright position but do not allow them to dry.

(iv)   First dip test slides into emulsion to make sure it is thoroughly melted and homogeneous. Slides should be uniformly coated with a thin grey film when held up to safe light.

(v)   Dip the experimental slides into emulsion. On withdrawal drain them against the edge of the cell for 5 sec, then turn them the right way up and place the labelled end in groove of the drying rack (i.e., slides should be in a vertical position).

(vi)   Slides will dry in 1 h if left alone, alternatively they can be dried with a fan for $5-10$ min.

(vii)   When dry, place them in light-tight black plastic boxes containing tubes of silica gel and place the boxes in a refrigerator at 4°C.

(viii) It is recommended to develop test slides at different intervals to find the most effective duration of exposure. The following times have been found effective and the results could be used as a guide: [^{3}H]thymidine, $6-7$ days; [^{3}H]uridine, $8-10$ days; [^{3}H]leucine, $8-10$ days.

10.4.3 *Development*

(i)   Use Kodak K19 developer.

(ii)   Filter solution before use.

(iii)   Develop for $4-5$ min at $18-21$°C.

(iv)   Rinse once with distilled water.

(v)   Fix in Amfix (1:10) for $2-3$ min or until film is clear.

(vi)   Rinse and place in hypoclearing agent (Kodak) for 2 min.

(vii)   Wash for 5 min in running tap water, then with three changes of deionised or distilled water.

10.4.4 *Staining*

Slides can be dried but better results are obtained if they are taken directly from distilled water to stain. Stain slides routinely with haemotoxylin-eosin or carmalum.

10.4.5 *Interpretation*

On microscopic examination, the label appears as dark grains over the nuclei ([^{3}H]thymidine, [^{3}H]uridine) or over the cytoplasm ([^{3}H]leucine). If the incubation with [^{3}H]uridine is prolonged beyond 40 min the label will have moved partially to the cytoplasm.

    As the film is very thin and homogenous, the histological structure of the tissues and changes of their growth pattern can be recognised and related to the localisation and degree of labelling.

The incorporation can be assessed quantitatively by counting the number of labelled and unlabelled cells and expressing the result as a percentage of total cell number and its standard deviation (labelling index). The uptake per cell is estimated by counting the number of grains over individual cells and expressed as average grain number per cell and its standard deviation (grain count). In general, a higher isotope concentration, and resultant grain density, should be used for labelling index measurements, so that the nucleus is totally black and easier to score, and a lower grain density $(20-50$ grains/cell) for grain counts.

11. GUIDE TO QUANTITATION OF RESULTS

Changes in morphology induced by chemical or physical means can be quantitated and alterations of morphology related to changes in cell proliferation or DNA, RNA or protein synthesis. The average number of explants used for each point should be six.

To eliminate possible variations of effect in different areas of the treated explants it is recommended to use serial sections and count either all or every second section.

For example, to quantitate the incidence of hyperplasia induced by a carcinogen or mitogen, hyperplastic and normal structures are counted and the result expressed as the percentage of hyperplastic structures and its standard error.

To determine the incidence of mitosis, dividing and resting cells are counted in serial sections of explants incubated with colcemid (2 μg/ml) or another metaphase-arresting agent, for 5 h prior to fixation. The results are expressed as the percentage of dividing cells over total cell number and its standard error.

DNA, RNA or protein synthesis can be determined in autoradiographs by counting labelled and unlabelled cells in serial sections of explants labelled with the appropriate isotopes. The result is expressed as the percentage of labelled cells over total cell number and its standard error.

This method has provided reliable data which were reproducible in replicate experiments. One or two replicates sufficed to confirm the results.

12. ADVANTAGES AND LIMITATIONS OF ORGAN CULTURE

In organ culture the various tissue components, their spatial relationship and functional activity are, under suitable conditions, preserved *in vitro*. The preservation of the stroma is of particular importance as it seems indispensable for the growth and differentiation of epithelium. These features make tissues in organ culture a more physiological experimental model than cell cultures. Contrary to some beliefs, the setting up of organ cultures does not require more effort than the establishment of cell cultures.

Foetal organs or organ rudiments develop *in vitro* in a similar manner as *in vivo*. Epithelia, in the presence of their stroma, grow and differentiate in the appropriate manner and manufacture specific secretory products. Hormone-dependent organs such as prostate and mammary glands, vagina and uterus remain hormone sensitive and responsive *in vitro*. Endocrine organs, such as ovaries, testes, adrenals and pituitary, continue to secrete specific hormones *in vitro*.

Thus, tissues in organ culture can provide important basic information on normal development, growth and differentiation and the influence of extrinsic factors on these parameters. The results can usually be obtained more quickly *in vitro* than *in vivo* and, if necessary, be easily quantitated. This quantitation has proved reliable and reproducible in replicate experiments.

The point has to be considered whether the use of organ culture can fully replace animal experimentation. *In vivo*, the interpretation of experimental results is complicated by the presence of systemic factors and their elimination *in vitro* reduces the problem to be investigated to its essentials. However, it is not certain whether the results obtained *in vitro* can always be extrapolated to the situation *in vivo*. The very presence of systemic factors may have a controlling influence and modify the response seen in culture. Studies on drug action sometimes produce contradictory results *in vitro* and *in vivo* because some drugs are metabolised *in vivo* but not *in vitro*.

The time factor poses another problem. The period of cultivation in organ culture is not unlimited and, so far, has not exceeded several months. This period may be too short where prolonged exposure to extrinsic factors is necessary. In such cases, a combined *in vitro-in vivo* procedure may be of benefit in which tissues treated *in vitro* are implanted into suitable host animals. The use of nude mice facilitates such procedures.

In summary, experiments in organ culture do not always replace animal experiments but the results seen *in vitro* serve as a valuable guide to the events taking place *in vivo* and thus considerably reduce the number of animal experiments necessary for studying any particular problem.

13. REFERENCES

1. Loeb,B. (1897) *Uber die Entstehung von Bindegewebe, Leucocyten und roten Blutkorperchen aus Epithel und eine Methode, isolierte Gewebsteile zu züchen*, published by M. Stern & Co, Chicago.
2. Harrison,R.G. (1907) *Proc. Soc. Exp. Biol. Med.*, **4**, 140.
3. Loeb,L. and Fleischer,M.S. (1919) *J. Med. Res.*, **40**, 509.
4. Parker,R.C. (1936) *Science (Wash.)*, **83**, 379.
5. Medawar,P.B. (1948) *Q. J. Microsc. Sci.*, **89**, 187.
6. Fell,H.B. and Robison,R. (1929) *Biochem. J.*, **23**, 767.
7. Lasnitzki,I. (1963) *Natl. Cancer Inst. Monogr.*, **12**, 281.
8. Rudnick,D. (1938) *J. Exp. Zool.*, **78**, 369.
9. Gaillard,P.J. (1951) in *Methods in Medical Research*, Vol. **4**, Visscher,M.B. (ed.), Year Book Publishers, Chicago.
10. Spratt,N.T., Jr. (1947) *Science (Wash.)*, **106**, 542.
11. Spratt,N.T., Jr. (1948) *J. Exp. Zool.*, **107**, 39.
12. Wolff,E.T. and Haffen,K. (1952) *Tex. Rep. Biol. Med.*, **10**, 463.
13. Chen,J.M. (1954) *Exp. Cell Res.*, **5**, 10.
14. Richter,K.M. (1958) *J. Oklahoma State Med. Assoc.*, 252.
15. Lash,J., Holtzer,S. and Holtzer,H. (1957) *Exp. Cell. Res.*, **13**, 292.
16. Shaffer,B.B. (1956) *Exp. Cell Res.*, **11**, 244.
17. Trowell,O.A. (1954) *Exp. Cell Res.*, **6**, 246.
18. Trowell,O.A. (1959) *Exp. Cell Res.*, **16**, 118.
19. Trowell,O.A. (1961) *Coll. Intern. CNRS*, **101**, p. 237.
20. Barret,L.A., McDowell,E.M., Frank,A.L., Harris,C.C. and Trump,B.F. (1976) *Cancer Res.*, **36**, 1003.
21. Hillman,E.A., Valerio,M.G., Halter,S.A., Barret-Boone,L.A. and Trump,B.F. (1983) *Cancer Res.*, **43**, 245.
22. Hillman,E.A., Vocci,M.J., Schurch,W., Harris,C.C. and Trump,B.F. (1980) in *Methods in Cell Biology*, Vol. **21**, Harris,C.C., Trump,B.G. and Stoner,G.D. (eds.), Academic Press, New York, p. 331.
23. Schurch,W., McDowell,E.M. and Trump,B.F. (1978) *Cancer Res.*, **38**, 3723.
24. Paul,J. (1975) *Cell and Tissue Culture*, Fifth Edition, published by E & S Livingstone, London.
25. Lasnitzki,I. (1951) *Br. J. Cancer*, **5**, 345.
26. Fell,H.B. and Canti,R.G. (1934) *Proc. R. Soc. Lond., Ser. B*, **119**, 470.
27. Fell,H.B. (1939) *Trans. R. Soc. Lond., Ser. B*, **229**, 407.

28. Fell,H.B. and Mellanby,E. (1953) *J. Physiol.*, **119**, 470.
29. Lasnitzki,I. (1963) *J. Exp. Med.*, **118**, 1.
30. Lasnitzki,I. (1955a) *J. Endocrinol.*, **12**, 236.
31. Lasnitzki,I. (1955b) *Br. J. Cancer*, **9**, 435.
32. Lasnitzki,I. (1958) *Br. J. Cancer*, **12**, 547.
33. Lasnitzki,I. (1956) *Br. J. Cancer*, **10**, 510.
34. Murray,M.R. and Stout,A.O. (1947) *Am. J. Anat.*, **80**, 225.
35. Hardy,M.H. (1978) *Am. Tissue Culture Assoc. Publ.*
36. Hardy,M.H. (1967) *Exp. Cell Res.*, **46**, 377.
37. Biggers,J.D., Claringbold,P.J. and Hardy,M.H. (1956) *J. Physiol.*, **131**, 497.
38. Moon,Y.S. and Hardy,M.H. (1973) *Am. J. Anat.*, **138**, 253.
39. Wolff,Et., Haffen,K., Kieny,M. and Wolff,Em. (1953) *J. Embryol. Exp. Morphol.*, **1**, 55.
40. Wolff,Et. (1960) *C.R. Hebd. Seances Acad. Sci. Paris,* **250**, 3881.
41. Wolff,Et., Smith,J. and Wolff,Em. (1975) in *Organ Culture in Biomedical Research*, Balls,M. and Monnickendamm,M.A. (eds.), Cambridge University Press, Cambridge, p. 405.
42. Lasnitzki,I. and Mizuno,T. (1977) *J. Endocrinol.*, **74**, 47.
43. Lasnitzki,I. and Misuno,T. (1979) *J. Endocrinol.*, **82**, 171.
44. New,D.A.T., Coppola,P.T. and Cockcroft,D.L. (1976) *J. Reprod. Fertil.*, **48**, 219.
45. New,D.A.T. (1983) in *Methods for Assessing the Effects of Chemicals on Reproductive Functions*, Vouk,V.B. and Sheehan,F.J. (eds.), p. 277.
46. Cockcroft,D.L. (1980) *Acta Morphol. Acad. Sci. Hung.*, **28**, 117.
47. Fainstat,T. (1968) *Fertil. Steril.*, **19**, 317.
48. Lasnitzki,I. (1976) in *Organ Culture in Biomedical Research*, Balls,M. and Monnickendamm,M.E. (eds.), Cambridge University Press, Cambridge, p. 241.
49. McMahon,M.J. and Thomas,G.H. (1973) *Br. J. Cancer*, **27**, 323.
50. Lasnitzki,I., Whitaker,R.H. and Withycombe,J.F.R. (1975) *Br. J. Cancer*, **32**, 168.
51. Hodges,G.M., Hicks,R.M. and Spacey,G.D. (1977) *Cancer Res.*, **37**, 3720.
52. Knowles,M.A., Finesilver,A., Harvey,A.E., Berry,R.Z. and Hicks,R.M. (1983) *Cancer Res.*, **43**, 374.
53. Lasnitzki,I. (1968) *Cancer Res.*, **28**, 510.
54. Lasnitzki,I. and Goodman,D.S. (1974) *Cancer Res.*, **34**, 1564.
55. Lasnitzki,I. (1976) *Br. J. Cancer*, **34**, 239.
56. Lasnitzki,I. and Bollag,W. (1982) *Cancer Treat. Rep.*, **66**, 1375.
57. Lasnitzki,I. Unpublished observations.
58. Doniach,I. and Pelc,S.R. (1950) *Br. J. Radiol.*, **23**, 184.

CHAPTER 8

Cytotoxicity and Viability Assays

ANNE P.WILSON

1. INTRODUCTION

Drug development programmes for the identification of new cancer chemotherapeutic agents involve extensive pre-clinical evaluation of vast numbers of chemicals for detection of anti-neoplastic activity. The safety evaluation of compounds such as drugs, cosmetics, food additives, pesticides and industrial chemicals necessitates the screening of even greater numbers of chemicals. Animal models have always played an important role in both contexts, and although cell culture systems have figured largely in the field of cancer chemotherapy, where the potential value of such systems for cytotoxicity and viability testing is now widely accepted, there is increasing pressure for a more comprehensive adoption of *in vitro* testing in both spheres of application. The impetus for change originates firstly from financial considerations, since *in vitro* testing has considerable economic advantages over *in vivo* testing. Secondly, it occurs from a realisation of the limitations of animal models in relation to man as increasing numbers of metabolic differences between species come to be identified. Finally, it occurs from the moralistic viewpoint in terms of reducing animal experimentation.

The safety evaluation of chemicals involves an extensive range of studies on mutagenicity, carcinogenicity, teratogenicity and chronic toxicity, all of which are outside the scope of this text, but an obvious area of overlap between cancer chemotherapy and safety evaluation is that of acute toxicity. A precedent has been set with the volume of research carried out in the last 40 years in cancer chemotherapy, and those systems which have been found to be most relevant in this field must be of interest both in general toxicity and anti-neoplastic activity.

The fundamental requirements for both applications are similar. Firstly, the assay system should give a reproducible dose-reponse curve with low inherent variability over a concentration range which includes the *in vivo* exposure dose. Secondly, the selected response criterion should show a linear relationship with cell number and thirdly, the information obtained from the dose-response curve should relate predictively to the *in vivo* effect of the same drug.

2. BACKGROUND

Use of *in vitro* assay systems for the screening of potential anti-cancer agents has been prevalent since the inception of clinical cancer chemotherapy in 1946, following the discovery of the anti-neoplastic activity of nitrogen mustard. A number of reviews describe the historical development of these techniques and their application (1,2) and more recent publications update to the present situation (3 – 5). The situation pertaining in the field of safety evaluation of drugs has also been reviewed in the publication of a recent symposium convened by the Humane Research Trust (6).

Early cytotoxicity studies were largely qualitative, in that explant cultures growing in undefined medium were used for the study of drug effect, which could be 'quantitated' by assessment of either morphological damage or of inhibition of the zone of outgrowth. The development of a semi-defined growth medium, together with techniques for growing dispersed cells as a monolayer on glass, allowed the screening of identical replicate cell samples in reproducible growth conditions, which therefore meant that drug effects could be quantitated in a meaningful way. Using these techniques in conjunction with measurement of protein content of treated and untreated cells, Eagle and Foley were able to demonstrate a clear-cut correlation between *in vitro* and *in vivo* activity of neoplastic agents (7) demonstrating the validity of the method. The assay system was subsequently included in the Cancer Chemotherapy National Service Centre (CCNSC) screening programme and currently plays an essential role in pre-screening during the step-wise purification of natural fermentation beers, because its rapidity, minimal requirement for test compound and economy offer advantages which cannot be found in the *in vivo* screening systems (3).

Accumulated experience, both clinically and experimentally, demonstrated that heterogeneity of chemosensitivity existed between tumours, even those of identical histology. The successful development of the *in vitro* agar plate assay for antibiotic screening precipitated interest in the development of an analogous technique for 'tailoring' chemotherapy to suit the individual tumour and patient, thus removing the undesirable combination of ineffective chemotherapy in the presence of non-specific toxicity. The idea was first applied experimentally by Wright *et al.* (8) using explants of human tumor tissue and this report has been succeeded by numerous others, which aim to investigate the correlation between *in vitro* and *in vivo* results in humans. The methodology which has been used is varied and represents the multiplicity of factors which must be considered in devising these assays. In spite of this diversity the consensus of the majority of reports is that more than 90% positive correlation can be expected between *in vitro* resistance and clinical resistance, and approximately 60% positive correlation between *in vitro* sensitivity and clinical response. There is also an indication that the frequency of responding cultures *in vitro* is similar to the frequency of responding tumours *in vivo* (9,10).

The relationship between *in vitro* drug sensitivity exhibited by primary cultures of human tumours and their *in vivo* counterparts argues for their use in drug evaluation programmes since they provide a closer approximation to the human clinical situation than do the limited number of cell lines which are currently used. The role of the 'Human Tumour Stem Cell Assay' is under investigation in this context (11).

In vitro culture systems are also used extensively for mechanistic studies on drug action for which the same technical considerations apply, although more specific biochemical end-points may be required.

3. SPECIFIC TECHNIQUES

Decision making on the final choice of assay is a function of the context in which the assay is to be used, the origin of the target cells and the nature of the test compound.

Parameters which vary between different assays include:

(i)   culture method;
(ii)   duration of drug exposure;
(iii)   duration of recovery period after drug exposure;
(iv)   end-point used to quantitate drug effect.

3.1 Culture Methods

The choice of culture method depends on the origin of the target cells and the duration of the assay and, to some extent, dictates the end-point.

3.1.1 *Organ Culture*

The advantages offered by organ culture relate to the maintenance of tissue integrity and cell-cell relationships *in vitro*, thereby giving a closer analogy to the *in vivo* situation than the majority of other culture methods available. Reliable quantitation of drug effect is not facilitated by difficulties associated with size variation between replicates and cellular heterogeneity. Although the method has been used extensively to study the hormone sensitivity of potentially responsive target tissues, the number of studies relating to drug sensitivity is limited. The topic has been covered in a recent review article (12) (see also Chapter 7).

3.1.2 *Spheroids*

Spheroids result from the spontaneous aggregation of cells into small spherical masses, which grow by proliferation of the component cells. Their structure is analogous to that of a small tumour nodule, and the use of spheroids for drug sensitivity testing therefore permits an *in vitro* analysis of the effects of three-dimensional relationships on drug sensitivity, without the disadvantages previously mentioned for organ culture. Specific parameters which can be studied are:

(i)   drug penetration barriers in avascular areas;
(ii)   the effects of metabolic gradients (e.g., pO_2, pCO_2);
(iii)   the effects of proliferation gradients.

The majority of studies have been carried out using spheroids derived from cell lines, but primary human tumours also have the capacity to form spheroids in approximately 50% of cases; the spheroid-forming capacity of normal cells is limited in comparison with tumour cells which is an important consideration since stromal elements may be excluded during reaggregation of human tumour biopsy material. Culture times in excess of 2 weeks are usually necessary for drug sensitivity testing, and the method is not suitable therefore in a situation where results are required quickly.

3.1.3 *Suspension Cultures*

(i) *Short-term cultures (4 — 24 h).* The short-term maintenance of cells in suspension for assay of drug sensitivity is applicable to all cell sources. When the cells are derived from human tumour biopsy material the assay system has several theoretical advantages in that the ability to grow is not a limiting factor since no growth is required, stromal cell overgrowth and clonal selection are minimised and results can be obtained rapidly, which is important if the assay is to be used in a clinical context. The method has been used extensively in West Germany for chemosensitivity studies on a variety of tumour types (13,14). A modified method using either tissue fragments or cells has been described by Silvestrini *et al.* (10) again with a variety of tumour types. Both groups used the incorporation of tritiated nucleotides into DNA/RNA as an end-point. Limitations of the method relate mainly to the short time period of the assay which precludes long drug exposures over one or more cell cycles and also takes no account of either the reversibility of the drug's effect or of delayed cytotoxicity. The type of drug which can be tested is therefore theoretically restricted, and although one group have found the main value of the test to lie with adriamycin (14), the successful use of other drugs with differing modes of action has been reported (10).

(ii) *Intermediate duration (4 — 7 days)*. Suspension cultures of intermediate duration are particularly suited to chemosensitivity studies on haematological malignancies and have been described in several recent reports (15 — 17).

3.1.4 *Monolayer Culture*

The technique of growing cells as a monolayer has been most frequently applied to the cytotoxicity testing of cell lines, but the method has also been used with some success for studies on the chemosensitivities of biopsies from a variety of different tumour types. In the case of human biopsy material the greatest problems associated with the method are, firstly, that the success rate is limited because adherence and proliferation of tumour cells is not always obtained and, secondly, that contamination of tumour cell cultures by stromal cells (fibroblastic or mesothelial) is an all too frequent occurrence. The problem is greater with some tumour types than others (e.g., high for carcinoma of the breast, moderate for ovarian tumours and minimal for gliomas).

Some method of cell identification is therefore an essential part of such assays. Stromal cells appear to be more resistant to chemotherapeutic agents, but deliberate contamination of an ovarian tumour line (OAW 42) with up to 30% stromal cells had a minor effect on the chemosensitivity of the culture, implying that the majority of the measured response derived from the tumour cells. Heavier contamination with stromal cells gave intermediate values of sensitivity approximately as predicted. There was no evidence for the stromal cells conferring resistance on the tumour cells (*Figure 1*; Wilson, unpublished results). Similar results were obtained using drugs selected from antibiotics, anti-mitotics, alkylating agents and anti-metabolites.

The culture method probably offers the greatest flexibility in terms of possible drug exposure and recovery conditions, and also in methods of quantitation of drug effect. Of all methods described, the growth of cells in monolayers requires the lowest cell numbers, and it is therefore amenable to microscale methodology which permits multiple drug screening over a wide concentration range, and also facilitates automation. When cell numbers are really sparse it is often feasible to culture the cells until sufficient numbers are available for assay, although reports on changes in chemosensitivity after subculture are conflicting. Two to three subcultures are probably acceptable and, indeed, subculture has been recommended because variability between replicates is reduced (18). When stromal cell contamination is unacceptably high, subculturing also offers the possibility of 'purifying' tumour cell cultures by differential enzyme treatment or physical cell separation (see Chapters 5 and 6).

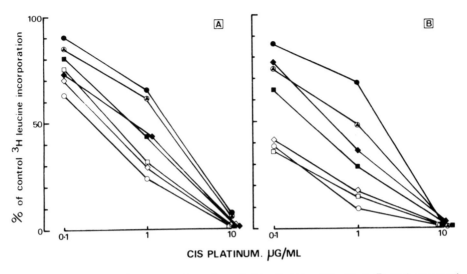

Figure 1. The effect of different proportions of mesothelial cells (**A**) and fibroblasts (**B**) on the measured sensitivity of an ovarian tumour cell line (OAW 42) to *cis*-platinum. (A.P.Wilson, unpublished results.) 10% stromal cells, □; 30% stromal cells, △;50% stromal cells, ■; 70% stromal cells, ♦ 90% stromal cells, ●; 100% OAW 42, ○; 100% stromal cells, ⦿. Chemosensitivity was determined using the microtitration plate assay described in Section 9.4.2.

3.1.5 *Clonogenic Growth in Soft Agar*

Although monolayer cloning can be applied to cells cultured directly from the tumour, the majority of reports in recent years have used suspension cloning to minimise growth of anchorage-dependent stromal cells. Clonogenic assays have the theoretical advantage that the response is measured in cells with a high capacity for self-renewal (potentially the stem cells of the tumour) and cells with limited proliferative capacity, which make up much of the bulk of the tumour, are not assayed. However, this is only true if colonies are truly clones (i.e., were initiated from one cell and not from a clump) and are only scored after many cell generations in clonal growth. Regrettably this is often not the case. Cloning efficiencies of $0.01-0.1\%$ are often quoted where it may be difficult to exclude the possibility of clumps, and while 10 generations (~ 1000 cells) may be readily obtained in monolayer cloning, suspension colonies are often scored after $4-6$ cell generations ($16-64$ cells). Given that some of these colonies started out as clumps of $3-4$ cells, the generation number may be as low as two and their capacity for self-renewal still be in some considerable doubt. Nevertheless, growth in soft agar is undoubtedly a useful assay system for cell lines which have a comparatively high plating efficiency. However, a number of technical problems have been encountered using solid tumours and effusions from patients which unfortunately influence interpretation of results. These include: difficulties in obtaining a pure single cell suspension from epithelial tumours (which is an essential requisite for the definition of clonogenic growth); very low plating efficiencies ($<1\%$); the formation of colonies from anchorage-dependent cells under certain growth conditions; requirements for large cell numbers; and finally the somewhat subjective nature of colony quantitation. These represent a failure of present technology rather than a failure of the assay method, and the importance

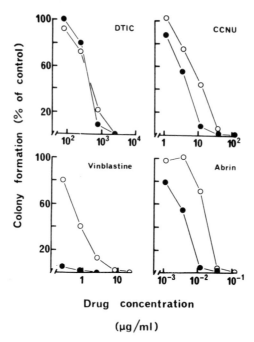

Figure 2. Dose-response curves of a melanoma xenograft (V.N.) cultured in soft agar using either (a) the 'Courtenay' method (○) or (b) the 'Hamburger-Salmon' method (●). Cells were exposed to the drugs for 1 h, plated at 3×10^4 cells per tube or dish and scored for colonies after 14 days incubation. Control cultures showed ~400 colonies using both methods. Method (b) showed greater sensitivity with three of the four drugs tested. (Reproduced with permission of the publishers, 26.)

of optimising methods of disaggregation and selective growth media for different tumour types has been emphasised (19). Critical assessment of the technical difficulties which have been encountered with the 'Human Tumour Stem Cell Assay' can be found in several reports (20 – 22). Results obtained using the double-agar method developed by Hamburger *et al.* (23) have been described in a recent review publication (24). An alternative methodology, developed by Courtenay and others (25), gives higher plating efficiencies, and has been compared with the 'Hamburger-Salmon' system by Tveit *et al.* (26), in which comparative study it was apparent that the methodology used influenced the chemosensitivity profile obtained (*Figure 2*).

4. DRUG CONCENTRATIONS

The choice of drug concentrations should be dictated by consideration of the therapeutic levels which can be achieved with clinically used drug dosages. When the compound is undergoing pre-clinical screening for potential activity this is not possible and, in the face of accumulated evidence on effective *in vitro* levels of compounds with known *in vivo* activity, an upper limit of 100 µg/ml is recommended. Pharmacokinetic data are available for many of the clinically used drugs, and parameters which are relevant to *in vitro* assays include the peak plasma concentration and the plasma clearance curves (*Figure 3*). A detailed description of pharmacological considerations may be found elsewhere (27). Pharmacokinetic data on most cancer chemotherapeutic agents, which

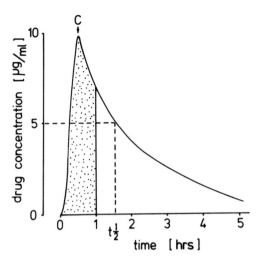

Figure 3. Typical plasma-clearance curve for intravenously administered drug. C = peak plasma concentration. t½ = terminal half-life of drug in plasma. ⊞ = area under curve for T = 1 h (μg/ml/h).

Table 1. A Comparison of Theoretical[a] Plasma Levels with Pharmacokinetic Data (28) for a Range of Cancer Chemotherapeutic Agents.

Drug	Dosage	Theoretical plasma level μg/ml	Measured C × T μg/ml/h
Melphalan	0.6 mg/kg	0.86	0.87 – 5.50
Cytosine arabinoside	10 mg/kg	14.29	15.23
BCNU	2.25 mg/kg	3.21	1.02
5-Fluorouracil	15 mg/kg	21.43	13 – 21
cis-platinum	100 mg/m²	3.8	1.42 – 2.52
	(e.g., S.A. = 1.6)		
Vinblastine	0.2 mg/kg	0.29	0.116 – 0.254

[a]The calculation is based on a 60 kg person with a 70% fluid compartment.

includes peak plasma concentration, the C × T parameter (where C = concentration and T = time in hours, and the terminal half-life (t½) of the drug in plasma has been summarised (28). When no pharmacokinetic data are available, an approximation of the plasma levels can be obtained by calculation of the theoretical concentration obtained when the administered dose is evenly distributed throughout the total body fluid compartment (*Table 1*). It is axiomatic that the concentration range adopted should give a dose-response curve, and the range selected by different groups reflects the different sensitivity levels of the various assays.

5. DURATION OF DRUG EXPOSURE

Pharmacokinetic data show that maximum exposure to drug occurs in the first hour after i.v. injection and, for this reason, an exposure period of 1 h has been chosen by many investigators. Whilst this may be adequate for cycle-specific drugs, such as the alkylating agents, longer exposure times over several cell cycles are necessary for phase-

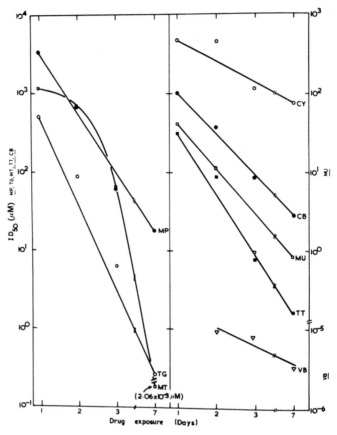

Figure 4. Effect of prolonged drug exposure on ID_{50} values for 6-mercaptopurine (MP), thioguanine (TG), methotrexate (MT), cyclophosphamide (CY), chlorambucil (CB), mustine (MU), thiotepa (TT) and vinblastine (BV), tested against HeLa cells using a microtitration plate assay (Section 9.4.2). Drugs were replaced at 24-h intervals for the 48-h and 72-h exposures, and at 24, 48 and 72 h for the 7-day exposure. A break in the time-axis is therefore shown between 3 and 7 days. (Reproduced with permission of the publishers, 29.)

specific drugs such as methotrexate and vinblastine. However, prolonged drug exposure using a variety of cancer chemotherapeutic agents has been shown to result in gradually decreasing ID_{50} values, as exposure times increase (29) (*Figure 4*). Rate of penetration of the drug may also be a limiting factor when short exposure times are used. The level of cell kill achieved with a short exposure time is also related to the method of assay; whilst high levels of cell kill can be obtained with a 1 h exposure to an alkylating agent using primary suspension cloning, a similar duration is insufficient to show cytotoxicity in monolayer.

Ultimately the question of duration of drug exposure becomes one of practicality. If a significant effect is achievable in 1 h, then this should be used. Many drugs may bind irreversibly to intracellular constituents and the actual exposure may therefore be in excess of 1 h due to drug retention. Others, principally the anti-metabolites and anti-tubulins, are more likely to be reversible if not present at the sensitive phase of

Table 2. Stability of Some Cancer Chemotherapy Drug Solutions at 37°C.

Drug	Stability data	Reference
Adriamycin	Stable during 48 h pre-incubation	
Bleomycin	Stable during 48 h pre-incubation	
5-Fluorouracil	Stable during 48 h pre-incubation	
Cytosine arabinoside	Stable during 48 h pre-incubation	A.P.Wilson (unpublished results)
cis-platinum	Full activity retained for 6 h, complete loss at 48 h	
Phosphoramide mustard	Full activity retained for 24 h, some loss at 48 h	
Melphalan	Full activity retained for 1 h, complete loss at 24 h.	
Chlorambucil	6-fold decrease in ID_{50} after 24 h pre-incubation	
Mustine	25-fold decrease in ID_{50} after 24 h pre-incubation	
Thiotepa	1.38-fold decrease in ID_{50} after 24 h pre-incubation	(29)
Cyclophosphamide	3.5-fold decrease in ID_{50} after 24 h pre-incubation	
Vinblastine	No change in ID_{50} after 24 h pre-incubation	

the cell cycle, and prolonged exposure spanning one or more cell cycles may be required. Resistance of the surviving fraction when short exposures are used may be due to an inappropriate phase of the cell cycle during drug exposure but truly resistant cells (i.e., with resistance even at the appropriate phase of the cell cycle) can only be demonstrated unequivocally after a prolonged exposure.

Regardless of the exposure time selected, it must be kept constant for each drug where tumours are being compared, and constant between drugs where two similar compounds are being compared. When prolonged drug exposure times are used, it should be remembered that the theoretical C × T value is only equal to the actual C × T value when the drug regains full activity at 37°C over the entire exposure period and the response is linear with time. Data on the stability of some drug solutions at 37°C are summarised in *Table 2*. The effective concentration of drug may also be reduced by binding of drug to the surface of the incubation vessel; this has been noted with adriamycin and actinomycin D.

6. RECOVERY PERIOD

The inclusion of a recovery period following drug exposure is important for three reasons:

(i)   when metabolic inhibition is used as an index of drug-effect; it allows recovery of metabolic perturbations which are unrelated to cell death;

(ii) sub-lethal damage can be repaired;

(iii) delayed cytotoxicity, such as occurs with 6-mercaptopurine and methotrexate, can be expressed.

Depending upon the nature of the drug and the end-point of the assay, absence of a recovery period can either under-estimate or over-estimate the level of cell kill achieved. However, it is equally important that the recovery period is not too long, because cell kill can be masked by overgrowth of a resistant population. In monolayer assays which monitor cell counts or precursor incorporation, the cells must remain in the log phase of growth throughout the exposure and recovery period. In clonogenic assays the recovery period is the period of clonal growth; the time taken to form measurable colonies is a minimum of five or six cell generations ($32 - 64$ cells/colony) in suspension assays and usually much greater than this when monolayer cloning is in use.

7. END-POINTS

7.1 Cytotoxicity, Viability and Survival

Interpretation of the significance of assay results is dependent upon distinguishing between assays which measure cytotoxicity and assays which measure cell survival. Cytotoxicity assays measure drug-induced alterations in metabolic pathways or structural integrity which may or may not be related directly to cell death, whereas survival assays measure the end-result of such metabolic perturbations which may be either cell recovery or cell death. Theoretically, the only reliable index of survival in proliferating cells is the demonstration of reproductive integrity, as evidenced by clonogenicity. Metabolic parameters also may be used as a measure of survival when the cell population has been allowed time for metabolic recovery following drug exposure.

When the test compound exhibits non-specific toxicity (i.e., toxicity which is not specifically related to proliferative potential) resulting in loss of one or more specific and essential cell functions rather than loss of reproductive capacity, a cytotoxicity test may be more appropriate.

7.2 Cytotoxicity and Viability

Some cytotoxicity assays offer instantaneous interpretation, such as the uptake of a dye by dead cells, or the release of ^{51}Cr or fluorescein from pre-labelled cells. These have been termed tests of viability and are intended to predict survival rather than measure it directly. On the whole these tests are good at identifying dead cells but may over-estimate long-term survival. Most imply a breakdown in membrane integrity and irreversible cell death.

Other aspects of cytotoxicity, measuring metabolic events, may be more accurately quantified and are very sensitive, but prediction of survival is less certain as many forms of metabolic inhibition may be reversible. In these cases impairment of survival can only be inferred if depressed rates of precursor incorporation into DNA, RNA or protein are maintained after the equivalent of several cell population doubling times has elapsed.

7.2.1 *Membrane Integrity*

This is the commonest measurement of cell viability at the time of assay. It will give an estimate of instantaneous damage (e.g., by cell freezing and thawing), or progressive damage over a few hours. Beyond this, quantitation may be difficult due to loss of dead cells by detachment and autolysis.

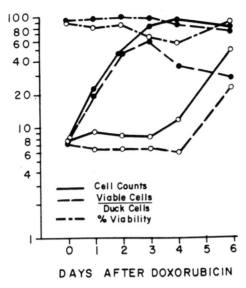

DAYS AFTER DOXORUBICIN

Figure 5. Effect of doxorubicin (0.12 μg/ml/h) on MDAY-D2 cells as assessed by three different techniques: Coulter counter particle counts (———) (Y axis units are cells/ml x 10^{-4}); ratio of living tumour cells to duck red blood cells, normalised to the same scale as the Coulter counts (- - - -), and % viability (— - —) Y axis units are % viability (living cells/living and dead cells x 100%). ● = control cultures; ○ = doxorubicin-treated cultures. (Reproduced with permission of the publishers, 35.)

(i) ^{51}Cr *release.* Labelling cells with ^{51}Cr results in covalent binding of chromate to basic amino acids of intracellular proteins. These labelled proteins leak out of the cell when the membrane is damaged, at a rate which is proportional to the amount of damage incurred. The method is used extensively in immunological studies for determining cytotoxic T cell activity against tumour target cells. Natural leakage of ^{51}Cr from undamaged cells is high and therefore the time period over which the assay can be used is restricted to approximately 4 h. In one comparative study which evaluated ^{51}Cr release as an end-point for drug cytotoxicity testing the method was found to be of no value (34).

(ii) *Dye exclusion.* Viability dyes which have been used to determine membrane integrity include trypan blue, eosin Y, naphthalene black, nigrosin (green), erythrosin B and fast green. Staining for viability assessment is more suited to suspension cultures than monolayers, because dead cells detach from the monolayer and are therefore lost from the assay. A major disadvantage is the failure of reproductively dead cells to take up dye, as was demonstrated when cells with impaired clonogenicity showed 100% viability according to dye exclusion (34). The method has been renovated however, and technical innovations introduced which attempt to circumvent some of the problems commonly associated with such assays. In the methodology developed by Weisenthal *et al.* (35), a 4-day assay period is used to permit reproductively dead cells to lose their membrane integrity and the inaccuracies produced by either overgrowth of viable cells or lysis of dead cells is compensated for by incorporation of fixed duck erythrocytes as an internal standard. Comparison of cell counts *versus* % viability *versus* viable cell/duck cell ratio demonstrated the increased sensitivity of the latter method (*Figure 5*). The method has been applied with equal success to solid tumours, effusions and haematological malignancies.

7.2.2 *Respiration and Glycolysis*

Drug-induced changes in respiration (O_2 utilisation) and glycolysis (CO_2 production) have been measured using Warburg manometry, both parameters showing dose-related depression (31,32). Other authors have determined dehydrogenase activity by incorporating methylene blue into agar containing drug-treated cells, cell death being indicated by non-reduction of the dye (33). The latter method has the disadvantage of being non-quantitative, whilst the former, although quantitative, has not been widely adopted because the technical manipulations involved are extensive and unsuited to multiple screening.

7.2.3 *Radioisotope Incorporation*

Measurement of the incorporation of radiolabeled metabolites is a frequently used end-point for cytotoxicity assays of intermediate and short-term duration.

(i) *Nucleotides.* Measurement of [³H]thymidine incorporation into DNA and [³H]uridine incorporation into RNA are two of the most commonly used methods of quantitation of drug cytotoxicity (10, 13, 14, 16). In short-term assays, which do not include a recovery period, there are a number of disadvantages, all of which relate to a failure of [³H]thymidine incorporation to reflect the true DNA synthetic capacity of the cell. These are:

(i)   changes may relate to changes in size of the intracellular nucleotide pools rather than changes in DNA synthesis;

(ii)   some drugs such as 5-fluorouracil and methotrexate which inhibit pyrimidine biosynthesis (*de novo* pathway) cause increased uptake of [³H]thymidine due to a transfer to the 'salvage' pathway, which utilises pre-formed pyrimidines;

(iii)   continuation of DNA synthesis in the absence of [³H]thymidine incorporation can occur (36).

The choice of isotope appears, to some extent, to be dependent on the drug; for example, Volm *et al.* (13) have recommended [³H]uridine for adriamycin, [³H]thymidine for 4-hydroxperoxycyclophosphamide and [³H]deoxyuridine for 5-fluorouracil, whilst Silvestrini *et al.* (10) reported similar uptake of [³H]thymidine and uridine for adriamycin, 4-hydroperoxycyclophosphamide, cis-platinum and mitomycin C, but differing uptake for bleomycin, vincristine, actinomycin D and VP-16. The low labelling index of human tumours with resultant low levels of nucleotide incorporation in short-term assays necessitates the use of high cell densities which can restrict the number of drugs and range of concentrations tested when cell numbers are limited. Two different 'hybrid' techniques have recently been reported, which seek to combine the 'stromal cell inhibition' advantage offered by the soft agar culture system with the facilitated quantitation offered by the use of radioisotopes. Both assays are of intermediate duration (~4 days) and use [³H]thymidine incorporation into DNA as an end-point; in one method the cells are grown in liquid suspension over soft agar (37), whilst in the other the cells are incorporated in the soft agar (38).

Given that a homogeneous cell population is available, [³H]nucleotide incorporation can be used after an appropriate recovery period to measure survival or, in the presence of drug, to measure an anti-metabolic effect, but with the reservations expressed above.

(ii) [^{125}I]*Iododeoxyuridine ([^{125}I]UdR).* [^{125}I]UdR is a specific, stable label for newly synthesised DNA which is minimally re-utilised and can therefore be used over a 24-h period to measure the rate of DNA synthesis (39); quantitation is facilitated because the isotope is a gamma emitter. Disadvantages include its variable toxicity to different cell populations, which therefore means that more cells are required because [^{125}I]UdR must be used in low concentrations.

(iii) [^{32}P]*Phosphate (^{32}P).* The rate of release of ^{32}P into the medium from pre-labelled cells is a function of the cell type and is increased in damaged cells. This has been used as a measure of drug efficacy (40). The incorporation of ^{32}P into nucleotides has also been used as an index of drug cytotoxicity (41). Neither method has been routinely adopted.

(iv) [^{14}C]*Glucose.* Glucose incorporation is used as a cytotoxicity end-point because it is a precursor which is common to a number of biochemical pathways (42). The method has not been widely used.

(v) [^3H]*Amino acids.* Protein synthesis may be considered as an essential metabolic process without which the cell will not survive, and incorporation of amino acids into proteins has been used successfully as an index of cytotoxicity. The most extensive studies have utilised monolayers of cells growing in microtitration plates, using either incorporation of [^3H]leucine (29) measured by liquid scintillation counting, or [^{35}S]-methionine incorporation, measured using autofluorography (43).

7.3 Survival (Reproductive Integrity)

Survival assays give a direct measure of reproductive cell death by measuring cloning efficiency either in monolayer or in soft agar. The end-points which have been described in the previous section can also be used as an index of reproductive integrity when the design of the assay incorporates a recovery period. Increases in total protein, cell number or protein synthetic capacity have been taken to imply proliferative ability although interpretation is more difficult due to differential responses in elements of a heterogeneous cell population.

7.3.1 *Cloning in Monolayer*

Cells generally have a higher cloning efficiency in monolayer than they do in soft agar, and the method is frequently used for cell lines. Normal cells and tumour cells will form colonies in monolayer, and the method is not therefore applicable to tumour biopsy material, which commonly shows high levels of stromal cell contamination, unless criteria are available to discriminate between tumour and stroma. Feeder layers of irradiated or mitomycin C-treated cells can be used to increase plating efficiencies, and indeed small drug-resistant fractions are more likely to be detected in the improved culture conditions existing when feeder cells are used (44).

7.3.2 *Cloning in Soft Agar*

The advantages and disadvantages of this method have been discussed in Section 3.1.5.

7.3.3 *Spheroids*

Various methods can be used to quantitate the effect of drugs on spheroidal growth. These are:

(i)   relative changes in volume of treated and untreated spheroids;
(ii)   cloning efficiency in soft agar of disaggregated spheroids;
(iii)   cell proliferation from spheroids adherent to culture surfaces.

The first method is rather insensitive because spheroidal growth tends to plateau, and the second may be affected by difficulties with disaggregation to a single cell suspension and low plating efficiencies. The choice of end-point is largely a function of the individual characteristics of the particular spheroids under study.

7.3.4 *Cell Proliferation*

An increase in cell number in a proliferating cell line can be regarded as an index of normal behaviour. Growth curves may be determined and the doubling time during exponential growth derived. An increase in the doubling time is taken as an indication of cytotoxicity, but it must be stressed that this is a kinetic measurement averaged over the whole population and cannot distinguish between a reduced growth rate of all cells and an increase in cell loss at each cell generation. It must be emphasised that estimates of cytotoxicity based on cell growth in mass culture must utilise the whole growth curve, or they may be open to misinterpretation. If 50% of cells die at the start of the experiment, the growth rate of the residue, determined in log phase, may be the same, but will show a delay. In practice it is very difficult to distinguish between early cell loss and a prolonged log period where cells are simply adapting. For these reasons, cell growth rates must be taken only as a rough guide to cytotoxicity, and accurate measurements of cell survival and cell proliferation should be made by colony-forming efficiency (survival) and colony size (proliferation).

7.3.5 *Total Protein Content*

Protein content determination is a relatively simple method for estimating cell number. It is particularly suited to monolayer cultures, and has the advantage that washed samples can be stored refrigerated for some time before analysis without impairment of results, facilitating large-scale screening. Over-estimation of cell number may arise with some drugs which inhibit replication without inhibiting protein synthesis (e.g., 5-bromodeoxyuridine, methotrexate). An adaptation of the Lowry method has been specifically developed for monolayer cultures (30), using the Folin-Ciocalteau phenol reagent. An alternative method using amido black has also been recommended because the linear section of the standard curve extends over a wider concentration range (3). Assessment of cytotoxicity by this method requires the demonstration of an alteration in the accumulation of protein per culture against time, preferably taken at several points, or at one point after prolonged drug exposure and recovery, as described above.

8. ASSAY COMPARISONS

In spite of the diversity of methodologies used for cytotoxicity and viability testing, the same levels of correlation between *in vitro* sensitivity and *in vivo* response have

Table 3. A Summary of Studies Undertaken to Compare the Results of Clonogenic Assays with the Results of Cytotoxicity Assays.

Authors	Methods compared	Cells used	Findings
Roper and Drewinko (34)	^{51}Cr release; dye exclusion; labelling index; growth kinetics	T_1 lymphoma	Clonogenicity only reliable dose-dependent index of drug effect
Tveit et al. (26)	'Courtenay' versus 'Salmon' soft agar systems	Melanomas: xenografts and biopsies	Chemosensitivity results not comparable: 'Salmon' system shows greater sensitivity
Morgan et al. (18)	Monolayer cloning versus microtitration assay (Sections 9.4.2 and 9.4.3)	Human astro-cytoma	Good correlation: cloning more sensitive in detection of small resistant fractions
Wilson et al. (45)	Short-term biochemical assay (Section 9.4.4) versus microtitration assay	Cell lines (T13 and MCF 7)	Comparability dependent on drug and exposure time. When not directly comparable, cut-off points selected from training data gave comparable results
Weisenthal et al. (17)	Dye exclusion	Animal tumour cell lines	Qualitative agreement between dye exclusion and clonogenic assay
Friedman and Glaubiger (37)	[^{3}H]TdR incorporation in liquid suspension over soft agar	Human tumour biopsy material	~89% correlation between [^{3}H]TdR incorporation and 'Human Tumour Stem Cell Assay

been reported when the various methods have been used on human tumour biopsy material. A number of comparative studies have been undertaken, in which clonogenicity has been taken as the standard for comparison. A summary of the results is shown in *Table 3*. It would appear that appropriately designed cytotoxicity assays give results comparable to clonogenic assays.

9. TECHNICAL PROTOCOLS

9.1 Drugs and Drug Solutions

9.1.1 Drug Sources

Pharmaceutical preparations for i.v. administration frequently contain various additives which may themselves be cytotoxic. Such preparations are therefore not suitable for aliquotting by weight and, if they are used as a stock solution, the cytotoxicity of the additional components should be determined in the assay system used. The problem can be avoided by obtaining pure compounds from the drug manufacturers.

9.1.2 Storage

It is recommended that dry compounds be stored at $-20°C$ to $-70°C$ over desiccant; this is especially important with compounds which are unstable in aqueous solution. It is routine practice to make up stock drug solutions which are then aliquotted and

Table 4. Solvents Used for Cancer Chemotherapy Drugs.

Soluble in aqueous solution	Insoluble	Solvents used
Adriamcyin	Chlorambucil	2% HCl:98% ethanol diluted with 4.5 vols of propane 1,2-diol and 4.5 vols saline or DMSO[a]
Actinomycin D		
Bleomycin		
Mitomycin C		
Cytosine arabinoside		
5-Fluorouracil		
Methotrexate	Melphalan	0.1 M HCl or DMSO
Vinblastine		
Vincristine		
Thiotepa	BCNU	Ethanol
cis-platinum		
4-Hydroperoxycyclophosphamide	CCNU	Ethanol plus 1 vol each of 5% Tween 80 and PBS[b]
Procarbazine	Hexamethylmelamine	DMSO or homogenise in DMSO plus 9 vols 5% Tween 80 in saline
	6-Mercaptopurine	DMSO

[a]Dimethylsulphoxide.
[b]Phosphate buffered saline.

stored at $-70°$C. Storage at a higher temperature is not recommended, and some drugs (nitrosoureas) are unstable even under these conditions (46). As some drugs may bind to conventional cellulose nitrate or acetate filters, sterile filtration must be carried out under controlled conditions to check for binding of drug. Use the maximum drug concentration and maximum volume to saturate the filter within the first few ml of filtrate, which may then be discarded. Alternatively use non-absorbent filters, e.g. nylon. In either case the filtrate should be assayed to make sure no activity has been lost. In practice it has been found that handling of non-pharmaceutical preparations of pure drugs under aseptic conditions (i.e., use of sterile blade for weighing out into sterile container) is sufficient to prevent contamination, even in experiments of prolonged duration.

9.1.3 Diluents

(i) *Solvents.* Solvents which can be used for the common cancer chemotherapeutic agents are shown in *Table 4*. Ethanol should be diluted out 1000-fold to avoid cytotoxicity; dimethyl sulphoxide is not normally cytotoxic at 1/100 dilution, but some primary cultures of human tumours may show exquisite sensitivity to the compound. Solvent controls should therefore be routinely included when primary culture material is being assayed.

(ii) *Medium components.* Certain drugs bind avidly to serum proteins (e.g., *cis*-platinum), and the presence of serum in the incubation medium may therefore reduce the amount of available drug. When the sensitivity of cells to *cis*-platinum was determined using a 0, 5, 10 or 20% serum concentration, the influence of serum was found to depend on the exposure time (*Figures 6A* and *B*). Thus, for a 3-h exposure, cells were more

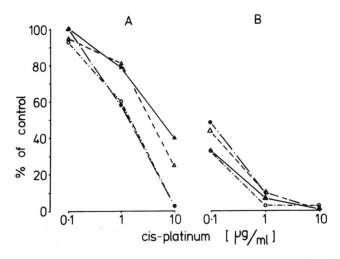

Figure 6. Effect of serum concentration on the sensitivity of an ovarian tumour cell line (OAW 42) to *cis*-platinum, measured using a microtitration plate assay (Section 9.4.2). Exponentially growing cells were exposed to *cis*-platinum for 3 h with a 72-h recovery period (**A**) or for 48 h with a 24-h recovery period (**B**). 0% serum, ○; 5% serum, ●; 10% serum, △; 20% serum, ▲. Standard deviations were ≤ ±10%. (A.P.Wilson, unpublished results.)

sensitive in the presence of low serum concentrations whilst consistent sensitivity in all serum concentrations was observed with a 48-h exposure period (A.P.Wilson, unpublished results).

Components of some media can protect cells against the cytotoxic effects of anti-metabolites (e.g., thymidine and hypoxanthine protect against methotrexate, thymidine protects against 5-bromodeoxyuridine and 5-FdUR). A comprehensive list has been detailed elsewhere (3). Such considerations are of particular importance when screening new agents for potential cytotoxic activity.

9.1.4 *Drug Activation*

Many compounds which are not themselves cytotoxic are converted to cytotoxic metabolites by the P450 mixed oxidase system of the liver. Cyclophosphamide is the best known of the cancer chemotherapeutic compounds to require *in vivo* activation, and methods of obtaining active metabolites have included use of urine from drug-treated patients, and also S9 microsome fractions prepared from the livers of phenobarbitone treated rats for *in vitro* activation. Pure preparations of phosphoramide mustard and 4-hydroperoxycyclophosphamide can now be obtained, and although the argument as to which of these compounds is responsible for *in vivo* anti-neoplastic activity is not fully resolved, experimental evidence seems to favour the 4-hydroperoxycyclo-phosphamide derivative (47).

Alternative methods for *in vitro* activation are under investigation using cultured rat hepatocytes or human liver biopsy material. The use of intact cells, in which the levels of co-factors resemble those *in vivo* and are high enough to sustain the associated reactions, provides a closer approximation to the *in vivo* situation, and indeed may give different results to those obtained using liver homogenates. Species differences exist in the complement of cytochrome-P450 system found in the liver, and for this reason the use of human liver homogenate may again provide a closer analogy to the *in vivo* situation, especially since individual variations in metabolic activity also exist. The subject has been discussed in detail elsewhere (48,49).

9.2 Drug Incubation

It is common procedure to incubate cells with drug solutions immediately after enzyme disaggregation of solid tissue, or harvesting of cell monolayers by trypsinisation. There is evidence to suggest that susceptibilty of cells to drug is altered by enzyme treatment and does not return to 'untrypsinised' levels until approximately 12 h after enzyme exposure (50). It may therefore be expedient to include a pre-incubation recovery period for freshly disaggregated cells to allow for this.

Maintenance of pH at 7.4 is essential during the incubation period since alterations in pH will alter cell growth and alkaline pH particularly will markedly reduce cell viability.

Comparison of static *versus* non-static incubations of cell suspensions revealed marked differences in dose-response profiles (51), and cells should therefore be kept in continuous suspension to allow equal drug distribution. If the surface area:depth ratio of the incubation vessel is small, incubation in a water-shaker bath will not keep the cells in suspension, and intermittent shaking by hand (e.g., at 10-min intervals) is recommended.

9.3 Assay by Survival and Proliferative Capacity

9.3.1 *Clonogenicity*

One of the most generally accepted methods for assaying for survival is the measurement of the ability of single cells to form colonies in isolation. This is usually achieved by simple dilution of a single cell suspension, and survival determined by counting the colonies which form. A lower threshold must be set in line with the doubling time of the cells being studied and the total duration of the assay. Five or six doublings (32 or 64 cells/colony) is usually taken as the lower threshold.

As some drugs may have an effect on cell proliferation as well as, or instead of, survival *per se*, it may be necessary to do a colony size analysis as well. This may be done by counting the number of cells per colony (very tedious and only possible in small colonies), by measuring the diameter (prone to error if cell size or degree of piling up changes) or by measuring absorbance of colonies stained with 1% crystal violet.

(i) *Monolayer cloning.* Adherent cells are plated onto a flat surface of glass or tissue culture-treated plastic, allowed to grow and form colonies, stained and counted. Drug treatment is best performed before subculture for plating, where highly toxic substances are being tested. For low grade toxins, where chronic application is required, they may be applied 24 – 48 h after plating, and retained throughout the clonal growth period, provided they are stable.

(1)   Prepare replicate 25 cm² flasks, two for each intended concentration of drug, and two for controls.

(2)   When cultures are at the required stage of growth (usually mid-log phase but, in special circumstances, may be in the plateau phase of the growth cycle) add the drug to the test and solvent to the control for 1 h at 37°C.

(3)   Remove the drug, rinse the monolayer with phosphate-buffered saline (PBS), and prepare a single cell suspension by conventional trypsinisation (see Chapter 1).

(4)   Count the cells in each suspension and dilute to the appropriate cell concentration to give 100−200 colonies per 5 or 6 cm Petri dish. This figure depends on the plating efficiency of the cells and the effect of the drug (e.g., control plates with no drug, from a cell culture of a known plating efficiency of 20%, will require 500−1000 cells per dish, at the ID_{50} of the drug, will require 1000−2000 cells per dish and at the ID_{90}, 5000−10 000 cells per dish). A trial plating should be done first:

(a)   to determine the plating efficiency of the cells,

(b)   to determine approximately the ID_{50} and ID_{90} of the drug.

In practice it is more usual to set up dishes at two cell concentrations, one to give a satisfactory number of colonies at low drug concentrations and controls, and one for higher drug concentrations, with some overlap in the middle range. Experience will usually determine where this is likely to fall.

(5)   Plate out the appropriate number of cells per dish and place in a humid incubator at 37°C with 5% CO_2.

(6)   Grow until colonies form. For rapidly growing cells (15−24 h doubling time) this will take 7−10 days, for slower growing cells (36−48 h doubling time) 2−3 weeks are required. In general, for a survival assay the colonies should grow to 1000 cells or more on average (10 generations). As the colonies increase in size the growth rate (particularly of normal cells) will slow down as the colonies tend to grow from the edge, and the slower growing colonies will tend to catch up. Hence if cell proliferation (colony size) is the main parameter, clonal growth should be determined at shorter incubation times giving smaller colonies with a wider size distribution.

(7)   Rinse dishes with PBS, fix in methanol or glutaraldehyde and stain with 1% crystal violet, rinse in running tap water, distilled water and dry.

(8)   Count colonies above threshold and calculate as a fraction of control. Plot on a log scale against drug concentration.

(ii) *Clonogenicity in Soft Agar Using a Double Layer Agar System.* The following procedure utilises a 1-h drug exposure period, and gives four replicates per test condition.

(1)   Prepare 35 mm Petri dishes with a 1 ml base layer of 0.5% agar in growth medium by mixing 1% agar (melted by autoclaving or boiling) at 45°C with an equal volume of double strength medium at 45°C and dispensing 1 ml aliquots in a pre-heated pipette.

(2)   Prepare a single cell suspension of the target cell population (see Chapters 1 and 6), adjust the cell concentration in growth medium to give 20 times the final concentration desired at plating and store at 4°C until ready to use.

(3)   Prepare drug dilutions in growth medium and aliquot out 900 μl of each concentration into duplicate tissue-culture grade tubes, including control tubes contain-

ing growth medium only and appropriate solvent controls. Because of the time-span involved when plating out multiple drugs and concentrations, additional controls are recommended for plating out at the beginning, middle and end of setting up the experiment. The controls thus incorporate variability due to the time involved in setting up the assay.

(4)   Check that the stock cell suspension still comprises single cells, and add 100 μl to each of the prepared tubes.

(5)   Incubate the tubes at 37°C for 1 h (see Section 8.2).

(6)   At the end of the incubation period centrifuge the tubes (2 min, 100 g) and wash the cells in 5 ml of saline. Repeat once more, and resuspend the cell pellet in 2 ml of growth medium. Keep the cells on ice to maintain cell viability whilst plating out replicates.

(7)   Centrifuge the duplicate set of tubes for one drug concentration; remove medium and add 2 ml of warmed growth medium containing 0.3% agar to each tube. Needle gently to disperse cell aggregates and plate out 1 ml aliquots onto each of four prepared bases. The final plating out is most easily accomplished using a 1 ml micropipette (Finn pipette or equivalent): if ~2 mm is cut off the end of the tips this prevents problems due to blockage of the small aperture by solidified agar.

(8)   Put the dishes to solidify on a cooled, horizontal surface.

(9)   Repeat for all test conditions, including controls at appropriate intervals throughout.

(10) Incubate the plates in a humidified atmosphere of 95% air/5% CO_2.

(11) Score the plates for colonies when control colonies have reached a pre-determined size. This is usually more than 50 cells for cell lines, but a size of more than 30 or 20 cells has been used when the population has a slow growth rate and a low plating efficiency, as with human tumour biopsies.

(iii) *Modifications*

(1)   *The 'Courtenay' method.* The 'Courtenay' method for suspension cloning (51) utilises rat red blood cells as feeder cells, and a 5% O_2 tension. The procedure outlined above may be used, with appropriate modification of final plating conditions.

(2)   *Feeder cells.* A linear relationship between plating efficiency and plated cell numbers may not occur from low to high cell numbers. When cell lysis has occurred due to drug treatment, the reduction in cell numbers can reduce the plating efficiency disproportionately in relation to seeding density rather than reflecting the clonogenicity of the remaining cell population. This problem can be circumvented by incorporating homologous feeder cells in the assay which have either been lethally irradiated with a ^{60}Co or ^{137}Cs source, or treated for approximately 12 h with 2 $\mu g/10^6$ cells of mitomycin C. The radiation dose needs to be established for each cell type, (e.g., 2000 – 3000 rads for lymphocytes, 6000 rads for lymphoid cell lines.

(3)   *Use of (2-(p-iodophenyl)-3-(p-nitrophenyl)-5 phenyl tetrazolium chloride) (INT).* Viable cells reduce colourless tetrazolium salts to a water-insoluble coloured formozan product, and the reaction has been used to distinguish viable colonies

from degenerate clumps when scoring finally. It should be noted, however, that viable colonies may become degenerate due to nutrient deficiency which may cause misleading results. The stain is made by dissolving INT violet in buffered saline to a final concentration of 0.5 mg/ml; dissolution is slow and the stain needs to be prepared 24 h prior to use. Add $0.5 - 1$ ml to each 35 mm Petri dish and incubate overnight at 37°C. Viable colonies then stain a reddish-brown colour.

(4) *20% O_2 versus 5% O_2.* The lower oxygen tension, which more closely resembles physiological levels of oxygen, is recommended for clonogenic assays because it results in higher cloning efficiencies. There is evidence to suggest that it also modifies the chemosensitivity profile of the cell population (52), producing enhanced cytotoxicity.

9.3.3 *Spheroids*

The experimental protocols outlined below are based on techniques described in a recent publication (53). Three end-points may be used to determine the cytotoxic effect of drugs on spheroids; these are: (a) volume growth delay; (b) clonogenic growth; and (c) outgrowth as a cell monolayer. Pre-selection of similarly sized spheroids and drug incubation is a common starting point.

(i) Pre-select spheroids in the chosen size range. Sizes in the range $150 - 250$ μm are just visible to the naked eye and can be selected using a Pasteur pipette.

(ii) The method of incubation with drug depends to some extent on the exposure period to be used. For 1-h exposures, which have been utilised most commonly, incubate spheroids in agar-coated ($0.5 - 1\%$) Petri dishes or in glass universal containers. For longer exposure times, incubate the spheroids in a spinner vessel which will keep them in continuous suspension and prevent adherence to the vessel walls.

(iii) At the end of incubation rinse the spheroids in two to three 5-min washes of drug-free medium.

(i) *Volume growth delay.* Treated spheroids can be grown as a mass culture and the volume of individual randomly chosen spheroids determined at set time intervals. However, from a statistical view-point it is recommended that successive measurements are made on individually isolated spheroids placed in either 24-well or 96-well multidishes with agar-coated bases. The smaller wells can be used for spheroids up to 600 μm in diameter.

(1) Plate out $12 - 24$ spheroids per drug concentration, and 24 controls into individual wells. For a 24-well plate use 0.5 ml of 1% agar for the base and add the spheroid in 1 ml of growth medium.

(2) At $2 - 3$ day intervals measure two diameters (x,y) at right angles to each other, using an eye piece graticule in an inverted microscope. Calculate the size of the spheroid as 'mean diameter' $(\sqrt{x.y})$ or as 'volume'

$$\left[\frac{4}{3} \pi \left(\frac{\sqrt{x \cdot y}}{2} \right)^3 \right],$$

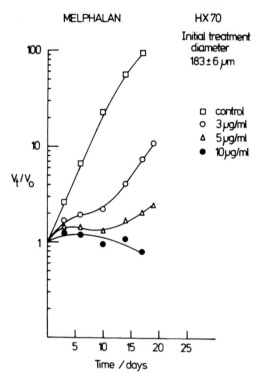

Figure 7. The effect of increasing concentrations of a 1-h exposure of melphalan on the growth of adenocarcinoma lung-derived spheroids. V_t = volume at time interval shown; V_0 = volume at start of experiment. Significant differences ($p < 0.001$: student's 't' test) were observed at day 14 between control and 3.0 μg/ml, and 3.0 μg/ml and 10.0 μg/ml. The diameter shown on the figure is the mean diameter ± s.e. of the plated spheroids. (Reproduced with permission of the publishers, 54.)

and use to construct growth curves for drug-treated and control spheroids. Results are usually normalised to pre-treatment spheroid size and expressed as V_t/V_0 where V_t = volume at time t and V_0 = initial volume. An example is shown in *Figure 7* for spheroids derived from the xenograft of an adenocarcinoma of the lung. The method assumes that the spheroid is symmetrical, but flattening of spheroids to a dome-shape has been observed for some cell lines, which will lead to over-estimation of volume.

(ii) *Clonogenic growth.*

(1) Disaggregate spheroids to a single cell suspension using appropriate conventional treatment with 0.25% trypsin, or 0.125% trypsin plus 1000 U/ml type I CLS grade collagenase (Worthington).

(2) Adjust the cell number to 20−200/ml for cell lines, or up to 10^3/ml for primary cloning. If cloning in suspension use 10^4/ml for cell lines and up to 5×10^5/ml for primary cloning.

(3) Score for colonies when the designated size criterion in controls has been attained.

(iii) *Outgrowth as cell monolayer.* Obtaining a pure single cell suspension may be virtually impossible, and plating efficiencies may be very low. The following method circumvents these problems.

(1) Allow individual spheroids to attach to individual wells of tissue culture plastic microtitration plates.
(2) After 2 – 3 weeks remove the central spheroid and determine the number of cells in the monolayer which has proliferated from the spheroid. This may be done directly using cell counts of the detached cell population. Alternatively, the methodology described for quantitation of drug effect on cell monolayers in Sections 9.4.2 and 9.4.3 could be adopted.

9.4 Cytotoxicity Assays

9.4.1 *Protein Determination for Cell Monolayers*

The following protocol is based on the procedure described by Oyama and Eagle (30), for test-tube cultures.

Reagents

Solution A	20 g sodium carbonate 4 g sodium hydroxide 0.2 g sodium potassium tartrate	in 1 litre distilled water
Solution B	0.5 g copper sulphate 100 ml distilled water	

Store A and B at 4°C.

Solution C	50 ml A + 1 ml B (make immediately before use)
Solution D	1 part Folin-Ciocalteau phenol reagent plus 1 part distilled water.

Serial dilutions of bovine serum albumin, 10 – 100 μg/ml, for construction of standard curve.

(i) Wash the cell sheet three times in PBS.
(ii) Invert the tubes and drain for 20 min.
(iii) Add 5 ml of solution C to each tube and leave at room temperature for 20 min.
(iv) Add 1 ml of distilled water to each test and blank and 1 ml of albumin to standards.
(v) Jet in 0.5 ml of solution D using a hypodermic needle and syringe, to ensure initial rapid mixing.
(vi) Mix the tubes on a rotamixer.
(vii) Read absorption on colorimeter at 660 mμ after 2 h. Include a reagent blank containing 5 ml of solution C, 1 ml of distilled water and 0.5 ml of solution D; because some protein binding from growth medium to glass/plastic occurs, the reagent blank should be prepared in tubes which have contained growth medium but no cells. The standard curve loses linearity at high protein concentrations and a dilution factor may therefore be needed when cell density is high. Results should be converted to protein concentration using the standard curve, before expressing drug-treated tubes as a percentage of the control.

This method may be adapted to multi-well plates reading absorbance in microcuvettes in a conventional spectrophotometer, or to 96-well microtitration plates reading absorbance on one of various plate readers available (Flow, Gibco, Ilacon).

9.4.2 Determination of Amino Acid Incorporation in Cell Monolayers, Growing in Microtitration Plates

The following protocol describes a cytotoxicity assay of intermediate duration and is based on a procedure first outlined by Freshney *et al.* (29). Exposure and recovery conditions may be varied to suit requirements.

(i)   Add 200 μl of cell suspension to each well of a 96-well microtitration plate. The cell density is dependent to some extent on the rate of proliferation of the cell type, and the length of the assay. It is desirable that confluence is not reached in control wells before the end of the assay, because their proliferation rate would then slow down relative to drug-treated wells, and confluent monolayers may detach from the well surface. For HeLa cells an initial concentration of 2×10^4/ml has been used, whilst for cell suspensions prepared from human ovarian tumour biopsies a higher concentration of $1-2 \times 10^5$/ml is recommended. Evaporation of medium from outer wells may occur, especially in longer assays, and it is therefore generally advised that the outer wells contain growth medium but no cells, in order to avoid the 'edge effect'. Edge effects due to evaporation can be reduced by using plate sealers (Mylar, Flow Laboratories). Some drugs are volatile, or give off volatile metabolites (e.g., formamides release formaldehyde), and this can cause variable non-specific cytotoxicity in adjacent wells. The use of plate sealers is essential for testing compounds such as these. Incubate the plates in a humidified atmosphere of air/5% CO_2.

(ii)   When the cells have attached to form a dividing monolayer (24 h for cell lines, but full adherence may take longer for primary and secondary cultures) replace the medium and dilute the drugs serially across the plate, using a minimum of three wells per drug concentration. Use one row for controls containing growth medium only.

(iii)   Incubate the plates with drugs for the chosen time period. If periods in excess of 24 h are to be used, add fresh drug solutions at 24-h intervals. Alternatively shorter exposures, e.g. 1 h, may be used, repeating daily if required.

(iv)   At the end of incubation, remove the drugs and gently wash the cell monolayers twice in PBS; add fresh medium (200 μl) to each well.

(v)   Incubate for the chosen recovery period; if this exceeds 3 days, change the medium at 3-day intervals, and check that the control wells have not attained confluence.

(vi)   At the end of the recovery period add 50 μl of $5-20$ μCi/ml [^{3}H]leucine (L-4,5 [^{3}H]leucine, Amersham plc) in growth medium to each well and incubate for 3 h.

(vii)   Remove the isotope and wash the cell monolayers three times in PBS. Care must be taken to avoid damaging the monolayers during pipetting procedures; this is best done by tilting the plate to an angle of 45°, and inserting the pipette tip in the angle between the side and base of the well.

(viii) Add 100 μl of PBS and 100 μl of methanol to each well, remove and add 100 μl of pure methanol to each well. Fix the cells for 30 min, which prevents detachment of the cell monolayer during subsequent processing.

(ix) Remove the methanol and air-dry the plates. Fixed plates can be stored for at least 48 h at 4°C before further processing.

(x) Put the plates onto ice and wash the monolayers in three 5 min washes of ice-cold 10% trichloroacetic acid (TCA). Following fixation, wash solutions can be removed by inversion of the plate and shaking sharply once or twice.

(xi) Wash off the TCA with methanol, air-dry the monolayer and add 100 μl of 1 N NaOH to each well; leave overnight at room temperature for the protein to solubilise.

(xii) Transfer the 100 μl of NaOH to individual minivials placed in glass scintillation vials.

(xiii) Add 2.4 ml of scintillant (e.g., Fisofluor, Fisons Ltd.) to each vial followed by 100 μl of 1.1 N HCl to acidify the contents.

(xiv) Cap the vials, mix to homogenise and clarify the contents, and count for 5 – 10 min on a β-counter.

(xv) Express the results as a percentage of control c.p.m.

Some cell populations obtained from human tumour biopsy material are poorly adherent, although they will proliferate in suspension. The protocol described is still feasible, providing that the plates are centrifuged at 1000 r.p.m. for 15 min prior to medium removal to avoid cell loss.

9.4.3 Measurement of Amino Acid Incorporation into Protein Using Autofluorography

The previous methodology using liquid scintillation counting is labour-intensive during processing, and can cause problems when many plates are processed, particularly when β-counting facilities are restricted. Minivials tend to lose scintillant through their walls after some days storage and, although this can be reduced by storing at 4°C prior to counting, it does mean that samples should be counted as soon as possible after preparation. An alternative method, which is well suited to automation, has been developed by Freshney et al. (43).

(i) Perform the procedures described in Section 9.4.2 (i) – (x) using 5 μCi/ml of [^{35}S]methionine (42 Ci/mol) to label the cells in step (vi); [^{3}H]leucine can be used but longer development times are required.

(ii) Add 50 μl of toluene-based scintillant to each well and centrifuge the open plates in microtitration plate carriers at 800 r.p.m. for 1 h at room temperature. Even evaporation of scintillant is obtained using this method which gives a flat layer of scintillant on the base of the well, and therefore improves resolution of spots. Sodium salicylate may be substituted for toluene-based scintillant, avoiding solubilisation of the plate and evolution of potentially toxic toluene vapour during evaporation.

(iii) Place a sheet of X-ray film under the plate in a dark room, and secure with a layer of polyurethane sponge and a pressure plate (metal or glass) using adhesive tape or bulldog clips. Expose the plates in a light-proof box at −70°C with desiccant. An exposure time of 5 days gives an absorbance of 0.92 when 3000 cells per well are exposed to 10 μCi/ml of [^{35}S]methionine (84 μCi/mol).

(iv)   Remove the film and develop for 5 min in Kodak D19 at 20°C. Fix in Ilfofix for 4 min, wash in Hypo clearing agent for 2 min, and tap water for 5 min. Dry the film.

(v)   The results are quantitated by scanning the images on a scanning densitometer (e.g., Chromoscan, Joyce Loebl, Gateshead, UK) with a thin-layer attachment using an 11 mm circular aperture and a blue (465 nm) filter. ID_{50} values can be obtained directly from the O.D. readings.

9.4.4 *Incorporation of [³H]Nucleotides into DNA/RNA*

The following procedure describes a short-term (3 h) cytotoxicity assay.

(i)   Adjust the cell concentration to 10 times the final desired viable cell concentration, usually $10^5 - 5 \times 10^5$ per ml.

(ii)   Titrate drug solutions at 900 μl per tube using replicates of three per concentration, including appropriate controls, and add 100 μl cell suspension per tube.

(iii)   Incubate the tubes at 37°C for 2 h (see Section 8.2) and add isotopes (2.5 μCi/ml) at the end of the second hour [e.g., 6-[³H]uridine (22 Ci/mmol); methyl-[³H]-thymidine (5 Ci/mmol); 6-[³H]deoxyuridine (25 Ci/mmol)]. Incubate for a further hour.

(iv)   Centrifuge the tubes and wash the cells three times in PBS to remove drug and isotope. It is essential that an adequate wash procedure is included, in order to exclude non-specific activity.

Further processing depends on available facilities. The simplest and most reliable method involves filtration of cells onto glass fibre filters under vacuum, followed by extraction with TCA, washing and drying. A range of devices are available including a single filtration unit which fits on a Buchner flask (Millipore UK, Ltd.), a multiple filtration head (Millipore UK, Ltd.) and various devices made for automated harvesting of microtitration plates, which come with adaptors for test tubes (Flow Laboratories, Dynatek, Ilacon). The single unit is not suitable for multiple samples because of the time involved in processing each sample. The basic procedure is as follows.

(i)   Harvest the cells onto glass-fibre filter paper, using 2 – 3 washes of the incubation vessel to ensure complete cell recovery.

(ii)   Wash the filters three times with 5 ml of ice-cold 10% TCA.

(iii)   Wash the filters sequentially with methanol, methanol-ether, and ether to remove water.

(iv)   Dry the filters and transfer them to scintillation vials with 10 ml of an emulsifer cocktail scintillant [e.g., Beckman Ready Solv HP (High Performance)].

(v)   Leave the samples for at least 1 h in the dark before counting to remove chemi-luminescence, which produces artefactually high counts.

(vi)   Count the samples for 5 – 10 min or as long as is required to reduce the counting error to ±5%.

Problems associated with the use of heterogeneous counting systems, as described above, have been detailed in an excellent technical review (55).

If filtration devices are not available, processing may be carried out in the incubation vessels, which should ideally be of glass.

(i)   Pellet the washed cells by centrifugation.

(ii)   Add 1 ml of ice-cold 10% TCA, resuspend the cells and leave them to precipitate for 5 min on ice. Precipitate the pellet and repeat the TCA extraction twice.

(iii) Wash the pellet in methanol, methanol-ether, and ether to remove water and dry.

(iv)   Add 1 ml of scintillation fluid to the residue which will dissolve, and transfer to scintillation vial. Make the volume up to 10 ml and count as before.

(i) *Modifications.* The following procedure has been described for measuring [³H]-thymidine incorporation into DNA in cells growing in liquid suspension over soft agar (37).

(i)   Add 25 μl of methyl-[³H]thymidine (6.7 Ci/mM diluted to 75 μCi/ml in PBS) to cells for 24 h.

(ii)   Harvest the cells and wash in PBS.

(iii) Add 5% TCA at 4°C to the washed cells for 30 min.

(iv)   Centrifuge and repeat (step 3) for 10 min.

(v)   Wash in methanol.

(vi)   Resuspend the pellet in 0.5 ml of 10 × hyamine hydroxide, and heat at 60°C for 1 h.

(vii) Place in 12 ml of Hydrofluor scintillation fluid and count.

9.4.5 *Total DNA Synthesis Measured by* [¹²⁵*I*]*UdR Uptake*

The methodology is based on descriptions in the literature (39).

(i)   Add 0.06 μCi/ml [¹²⁵I]UdR to growth medium of monolayer cultures for 24 h.

(ii)   Wash the cultures three times in saline and add fresh growth medium.

(iii) After 6 h, repeat the wash procedure to remove unbound label and label released from lysed dead cells, and count bound ¹²⁵I activity in a gamma counter.

As for the microtitration assay, confluence should be avoided in control tubes because overcrowding leads to reduced [¹²⁵I]UdR incorporation, and therefore underestimation of cell kill. When cells have been grown as monolayers in inclined test tubes the upper limit of the assay was found to be about 2×10^5 cells, and the lower limit about 2×10^4 cells.

10. INTERPRETATION OF RESULTS

10.1 Relationship Between Cell Number and Cytotoxicity Index

The validity of a cytotoxicity index is dependent upon the degree of linearity between cell number and the chosen cytotoxicity parameter, and this should be established for any cytotoxicity assay. In clonogenic assays a linear relationship may not occur at low cell numbers due to the dependence of clonogenic growth on conditioning factors, whilst at high cell densities linearity is lost due to nutritional deficiencies. In cytotoxicity assays linearity may be lost at the upper end due to density-dependent inhibition of the relevant metabolic pathway, whilst the sensitivity limit of the assay may affect linearity at the lower end. This would cause apparent stimulation at low drug levels and an overestimation of cell kill at higher concentrations. Control cell numbers at the end of the assay must therefore fall on the linear portion of the curve. The accuracy at high levels of cell kill is dependent upon the range over which linearity extends, and influences

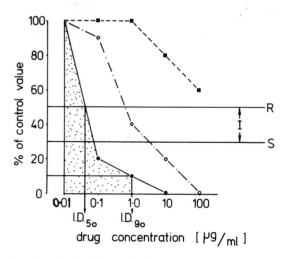

Figure 8. Interpretation of results using linear plot of response against drug concentration. ID_{50} and ID_{90} = concentration required to reduce cytotoxicity index to 50% or 90% of control value. ⊡ = area under curve between 0 and 1 μg/ml (for peak plasma concentration of 10 μg/ml). S = cut-off boundary for sensitivity; R = cut-off boundary for resistance; I = intermediae zone. ●——●, sensitive population; ○——○, intermediate population; ■——■, resistant population.

the number of decades of cell kill which can be measured. If results are plotted on a log-scale this implies that the assay is accurate down to $3-4$ decades of cell kill, which should be confirmed before expressing results in this way. This is particularly important in *in vitro* drug combination studies when synergism or additivity is often observed beyond the second decade of cell kill.

10.2 Dose-reponse Curves

Results are commonly plotted as dose-response curves using a linear scale for percentage inhibition (of isotope incorporation, for example) and a log scale for surviving fractions in clonogenic tests. Assay variation for replicate points is routinely depicted as mean ± standard deviation; a minimum of three replicates is therefore required for each test point. Some means is required for defining the sensitivity of a cell population in relation to other cell populations, or different test conditions: several parameters are available and are shown in *Figure 8*.

10.2.1 *Area Under Curve (AUC)*

The use of AUC recognises the probability that the shape of the dose-response curve may be instrumental in influencing the outcome to chemotherapy, rather than cell kill at any one concentration. It is calculated using the trapezoidal method which adds the area of rectangles and triangles under the survival curve. The method has been applied most extensively in the 'Human Tumour Stem Cell Assay' (56).

10.2.2 *Cut-off Points for Definition of Sensitivity and Resistance*

If plots of dose-response curves from multiple tumours show that tumours maintain their relative sensitivity rankings at different concentrations (i.e., crossing-over of dose-response curves is minimal), then information on the relative sensitivities of different tumours can be obtained by defining sensitivity at one concentration, and this is the most commonly used method for *in vitro* predictive testing. When retrospective correlations between *in vitro* data are made for defining these cut-off points, an intermediate zone is found where tumours cannot be defined as sensitive or resistant, and there is no clear-cut correlation between *in vitro* results and clinical response. The size of the intermediate zone will be at least partly related to the inherent variability of the assay, larger zones being associated with higher standard deviations. Although sensitivity may be defined at one concentration it is recommended that more than one concentration is tested, particularly in the developmental stage of an assay.

10.2.3 ID_{50} and ID_{90} Values

Tumour sensitivity may also be defined by the ID_{50} and ID_{90} values (i.e., drug concentration required to inhibit viability by 50% or 90%).

10.2.4 Correlation Between in Vitro and in Vivo Results

Criteria for defining tumours as sensitive or resistant are based on retrospective correlations between *in vitro* results and clinical responses, using a 'training set' of data. Even when a laboratory is using an established method for tumour sensitivity testing, 'own laboratory' sets of training data should be obtained to allow for inter-laboratory variation. The response of patients with tumours of intermediate sensitivity may be influenced by prognostic factors other than tumour sensitivity (e.g., tumour burden at onset of chemotherapy, stage of disease, histology, tumour cell doubling time, previous chemotherapy and performance status). When analysing results for correlations some attempt to stratify patients according to these parameters may assist in providing more meaningful data. Quantitative assessment of tumour response is also of paramount importance. It is pointed out that *in vitro* chemosensitivity can be expected only to indicate that *some* degree of cell kill will be achieved *in vivo*, not that the patient will achieve a complete response to treatment, the latter being under the influence of other factors also. The true positive correlation rate of an assay is defined as

$$\frac{S/S + S/R}{S/S} \times 100\%$$

where numerator of each fraction = *in vitro* response and demoninator = *in vivo* response. The true negative rate is defined as

$$\frac{R/R + R/S}{R/R} \times 100\%$$

In assessing the significance of the correlation rates obtained, these should be compared with the correlation rates which would be obtained were the *in vitro* results randomly distributed (57). For example, a drug gives a 50% response rate *in vivo*, and 50% of tumours show *in vitro* sensitivity to this drug. If the *in vitro* results are randomly distributed between sensitivity and resistance, then the chances of obtaining a positive correlation between *in vitro* sensitivity and *in vivo* response are 50% of 50% (i.e., 25%), and also of obtaining a positive correlation between *in vitro* resistance and *in vivo* resistance. The overall apparent positive correlation rate is therefore 50%. *In vitro versus in vivo* correlations are also complicated by the use of combination regimes to treat patients. Strictly speaking, correlations should be made only when *in vitro* data is available for all drugs used. Whether or not they are tested in combination depends on the treatment protocol since some drugs are administered sequentially. Also if the assay can only measure two decades of cell kill it may be too insensitive to detect additive or synergistic effects.

11. PITFALLS AND TROUBLE SHOOTING

Problems which may be encountered with these assays include:

(i)   large standard deviations;
(ii) variability between assays done on the same cell population;
(iii) stimulation to above control levels.

11.1 Large Standard Deviations

Possible reasons for large standard deviations include:

(i)   faults in aliquotting cell suspension, which are most likely to be due to inadequate mixing of cell suspension during dispensing leading to uneven distribution of cells between replicates;
(ii) the presence of large cell aggregates in the original cell suspension, leading to uneven distribution of cells between replicates;
(iii) non-specificity of cytotoxicity end-point [e.g., due to measurement of non-specific binding of radioactivity (see Section 9.4.4)].

11.2 Inter-assay Variation

Replicate assays on different days cannot be performed on human tumour biopsy material to check day to day reproducibility, but this can be evaluated using cell lines. It is a recognised problem that cell lines which show consistent sensitivity profiles may show 'deviant' results occasionally, for reasons which cannot be identified. Specific reasons for failure to obtain reproducible results may include:

(i)   failure to harvest the cell population at an identical time point (e.g., exponential growth *versus* early confluence *versus* late confluence);
(ii) deterioration of stock drug solutions (see Section 5);
(iii) when drug solutions have a short half-life they must be used immediately after diluting to ensure consistency in the drug levels available to cells in each assay;
(iv) failure to standardise incubation conditions (Section 9.2).

The assay system must be checked for reproducibility before applying it to human biopsy material.

11.3 Stimulation to Above Control Levels

Stimulation can be a true measure of cellular events but may be due to technical artefacts. These include:

(i)   non-specific binding of radioactivity;
(ii) density-dependent inhibition of metabolic pathways in controls which is not evident in test situations where some cell kill has been achieved;

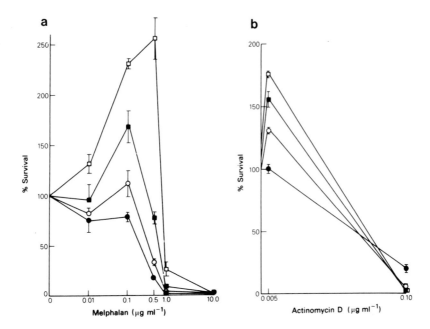

Figure 9. The effect of growth unit size on the survival curves of a murine melanoma cell line (CCL) to melphalan (**a**) and of a human melanoma biopsy to actinomycin D (**b**). Growth unit size and frequency was measured using FAS II automated image analysis system. ●, ≥ 60 μm; ○, ≥ 104 μm; ■, ≥ 124 μm; □, ≥ 149 μm. Mean ± s.e. shown. (Reproduced with permission of the publishers, 58.)

(iii)   stimulation of uptake of metabolic precursors by anti-metabolite (e.g., thymidine by 5-fluorouracil and methotrexate).

Recent results (58) have shown that plated cell density influences the distribution in size of growth units in clonogenic assays, with large units decreasing as plated cell numbers increase. The effect of this on drug sensitivity profiles was examined and, as expected, the dose-response curve was strongly influenced by the size criterion used for colony-scoring, with stimulation to above control levels occurring when large colonies were scored (*Figure 9*).

11.4 Cytotoxicity versus Anti-neoplastic Activity

When *in vitro* tests are being used as a pre-screen for anti-neoplastic activity it is important that a distinction can be made between non-specific cytotoxicity and specific anti-neoplastic activity since the *in vitro* test gives no measure of the therapeutic index of the drug. Guidelines for distinguishing between the two are suggested in *Table 5*, although it is realised that exceptions exist.

12. AUTOMATION

Automation is a necessary technical development for large-scale screening of drugs and tumours. Of the assays described none except possibly the microtitration plate assay can be automated during the manipulative procedures involved in setting up the assay. Cytotoxicity assays involving the use of radioiosotopes are the most readily quantitated

Table 5. Suggested Guidelines for Distinguishing Between Non-specific Cytotoxicity and Anti-neoplastic Activity.

Cytotoxicity	Anti-neoplastic activity
Equally effective against dividing and non-dividing cell populations	More effective against dividing cell populations than non-dividing cell populations
No heterogeneity of response between different tumour cell populations	Heterogeneity of response between different tumour cell populations
No difference in response of normal cells and tumour cells	Tumour cells more sensitive than normal cells
Primary biochemical lesions due to interference with cellular energy metabolism, or generation of free radicals	Primary biochemical lesions are: (i) inhibition of DNA replication or RNA transcription by binding to DNA; (ii) inhibition of nucleotide biosynthesis

automatically, though this may still require a high level of technical commitment which is alleviated by the use of the autofluorographic method described in Section 9.4.3. Various image analysers are also available for automated colony counting. Recently developed techniques for measuring the electrical impedance of cell cultures allows non-destructive determination of cytotoxicity at different time-points and is fully automated; such a technique has been used successfully to measure the drug sensitivity of a variety of cell populations (59).

13. FUTURE DEVELOPMENTS

In vitro cytotoxicity testing is a comparatively new concept in the field of safety evaluation and there are several aspects which merit investigation for evaluation of the role of these assays. These include:

(i)   development of defined bioactivation systems for metabolic conversion to cytotoxic metabolites;

(ii)   assay of agents with known cytotoxic activity for retrospective comparisons of *in vitro* effect;

(iii)   development of culture systems for growth of normal target cells;

(iv)   identification of the relevant *in vitro* cytotoxicity assay for compounds with differing modes of action, and intercomparison of different assay methods;

(v)   identification of mechanisms of cytotoxicity.

The increasing optimism regarding the role of *in vitro* cytotoxicity testing in cancer chemotherapy is reflected in the ever-increasing amount of literature on the subject, which has risen dramatically with the recent development of clonogenic assay systems for human biopsy material. An international conference was held in 1983 which provided comprehensive coverage of the present situation in *in vitro* predictive testing (60). Evidence for the retrospective accuracy of prediction is well-established, but the ultimate proof of the value of these assays awaits the results of a random prospective trial in which the response rates of previously untreated patients treated on the basis of either clinical choice or laboratory results are compared. Because of the large number of patients required for such a study, a multicentre trial is an essential requisite. Other aspects which need further development include:

(i)   identification of specific growth factors to increase the incidence of successful *in vitro* growth;

(ii)   establishment of more cell lines from different tumour types;

(iii) screening of new anti-neoplastic drugs and potentially active drugs (e.g., second generation agents), against cell lines of human tumour origin;

(iv) use of *in vitro* cytotoxicity testing for the design of drug combinations, evaluation of dose-dependency and optimum dose-scheduling;

(v)   isolation of resistant lines from treated patients for studies on mechanisms of developed resistance; and

(vi) cloning of cell lines from untreated patients for identification of heterogeneous drug sensitivity and mechanisms of primary resistance.

14. REFERENCES

1. Gellhorn,A. and Hirschberg,E. (1955) *Cancer Res.*, **15**, Suppl. 3, 1.
2. Foley,G.E. and Epstein,S.S. (1964) *Adv. Chemother.*, **1**, 175.
3. Hakala,M.T. and Rustrum,Y.M. (1979) in *Methods in Cancer Research: Cancer Drug Development. Part A*, DeVita,V.T. and Busch,H. (eds.), Academic Press, p. 247.
4. Dendy,P.P., ed. (1976) *Human Tumours in Short-term Culture: Techniques and Clinical Applications*, published by Academic Press.
5. Dendy,P.P. and Hill,B.T., eds. (1983) *Human Tumour Drug Sensitivity Testing in Vitro: Techniques and Clinical Applications*, published by Academic Press.
6. Turner,P., ed. (1983) *Animals in Scientific Research: An Effective Substitute for Man?*, published by MacMillan Press, Ltd.
7. Eagle,H. and Foley,G.E. (1956) *Am. J. Med.*, **21**, 739.
8. Wright,J.C., Cobb,J.P., Gumport,S.L., Golomb,F.M. and Safadi,D. (1957) *New Engl. J. Med.*, **257**, 1207.
9. Wilson,A.P. and Neal,F.E. (1981) *Br. J. Cancer*, **44**, 189.
10. Silvestrini,R., Sanfilippo,O. and Daidone,M.G. (1983) in *Human Tumour Drug Sensitivity Testing in Vitro*, Dendy,P.P. and Hill,B.T. (eds.), Academic Press, p. 281.
11. Salmon,S.E. (1980) in *Cloning of Human Tumour Stem Cells. Progress in Clinical and Biological Research*, Vol. **48**, Salmon,S.E. (ed.), Alan R. Liss Inc., NY, p. 281.
12. Masters,J.R.W. (1983) in *Human Tumour Drug Sensitivity Testing in Vitro*, Dendy,P.P. and Hill,B.T. (eds.), Academic Press, p. 163.
13. Volm,M., Wayss,K., Kaufmann,M. and Mattern,J. (1979) *Eur. J. Cancer*, **15**, 983.
14. KSST. Group for Sensitivity Testing of Tumours, *Cancer*, **48**, 2127.
15. Durkin,W.J,. Ghanta,V.K., Balch,C.M., Davis,D.W. and Hiramoto,R.N. (1979) *Cancer Res.*, **39**, 402.
16. Raich,P.C. (1978) *Lancet*, **i**, 74.
17. Weisenthal,L.M. and Marsden,J. (1981) *Proc. Am. Assoc. Cancer Res.*, **22**, 155.
18. Morgan,D., Freshney,R.I., Darling,J.L., Thomas,D.G.T. and Celik,F. (1983) *Br. J. Cancer*, **47**, 205.
19. Hill,B.T (1983) in *Human Tumour Drug Sensitivity Testing in vitro*, Dendy,P.P. and Hill,B.T. (eds.), Academic Press, p. 91.
20. Agrez,M.W., Kovach,J.S. and Lieber,M.M. (1982) *Br. J. Cancer*, **46**, 88.
21. Bertoncello,I. *et al.* (1982) *Br. J. Cancer*, **45**, 803.
22. Rupniak,H.T and Hill,B.T. (1980) *Cell Biol. Int. Rep.*, **4**, 479.
23. Hamburger,A.W., Salmon,S.E., Kim,M.B., Trent,J.M., Soehnlen,B., Alberts,D.S. and Schmidt,H.J. (1978) *Cancer Res.*, **38**, 3438.
24. Salmon,S.E., ed. (1980) *Cloning of Human Tumour Stem Cells. Progress in Clinical and Biological Research*, Vol. **48**, published by Alan R.Liss Inc., NY.
25. Courtenay,V.D. and Mills,J. (1978) *Br. J. Cancer*, **37**, 261.
26. Tveit,K.M., Endresen,L., Rugstad,H.E., Fodstad,Ø and Pihl,A. (1981) *Br. J. Cancer*, **44**, 539.
27. Alberts,D.S., George Chen,H.-S. and Salmon,S.E. (1980) in *Cloning of Human Tumor Stem Cells*, Salmon,S.E. (ed.), Alan R.Liss, Inc., NY, p. 197.
28. Alberts,D.S. and George Chen,H.-S. (1980) in *Cloning of Human Tumor Stem Cells*, Salmon.S.E. (ed.), Alan R.Liss, Inc, NY, Appendix 4.

29. Freshney,R.I., Paul,J. and Kane,I.M. (1975) *Br. J. Cancer,* **31**, 89.
30. Oyama,V.I. and Eagle,H. (1956) *Proc. Soc. Exp. Biol. Med.,* **91**, 305.
31. Bickis,I.J., Henderson,I.W.D. and Quastel,J.H. (1966) *Cancer,* **19**, 103.
32. Dickson,J.A. and Suzanger,M. (1976) in *Human Tumours in Short Term Cultures: Techniques and Clinical Applications,* Dendy,P.P. (ed.), Academic Press, p. 107.
33. Buskirk,H.H., Crim,J.A., Van Giessen,G.J. and Petering,H.G. (1973) *J. Natl. Cancer Inst.,* **51**, 135.
34. Roper,P.R. and Drewinko,B. (1976) *Cancer Res.,* **36**, 2182.
35. Weisenthal,L.M., Dill,P.L., Kurnick,N.B. and Lippman,M.E. (1983) *Cancer Res.,* **43**, 258.
36. Grünicke,H., Hirsch,F., Wolf,H., Bauer,V. and Kiefer,G. (1975) *Exp. Cell Res.,* **90**, 357.
37. Friedman,H.M. and Glaubiger,D.L. (1983) *Cancer Res.,* **42**, 4683.
38. Sondak,V.K. *et al.* (1984) *Cancer Res.,* **44**, 1725.
39. Dendy,P.P., Dawson,M.P.A., Warner,D.M.A. and Honess,D.J. (1976) in *Human Tumours in Short Term Culture: Techniques and Clinical Applications,* Dendy,P.P. (ed.), Academic Press, p. 139.
40. Forbes,I.J. (1963) *Aust. J. Exp. Biol.,* **41**, 255.
41. Izsak,F.Ch., Gotlieb-Stematsky,T., Eylan,E. and Gazith,A. (1968) *Eur. J. Cancer,* **4**, 375.
42. Edwards,A.J. and Rowlands,G.F. (1968) *Br. J. Surg.,* **55**, 687.
43. Freshney,R.I. and Morgan,D. (1978) *Cell Biol. Int. Rep.,* **2**, 375.
44. Freshney,R.I., Celik,F. and Morgan,D. (1983) in *The Control of Tumour Growth and its Biological Base,* Davis,W., Maltoni,C. and Tanneberger,St. (eds.).
45. Wilson,A.P., Ford,C.H.J., Newman,C.H. and Howell,A. (1984) *Br. J. Cancer,* **49**, 57.
46. Bosanquet,A.G. (1984) *Br. J. Cancer,* **49**, 385.
47. Powers,J.F. and Sladek,N.E. (1983) *Cancer Res.,* **43**, 1101.
48. Fry,J.R. (1983) in *Animals in Scientific Research: An Effective Substitute for Man?,* Turner,P. (ed.), Macmillan Press Ltd., p. 81.
49. Davies,D.S. and Boobis,A.R. (1983) in *Animals in Scientific Research: An Effective Substitute for Man?,* Turner,P. (ed.), Macmillan Press Ltd., p. 69.
50. Barranco,S.C., Bolton,W.E. and Novak,J.K. (1980) *J. Natl. Cancer Inst.,* **64**, 913.
51. Courtenay,V.D. and Mills,J. (1981) *Br. J. Cancer,* **44**, 306.
52. Gupta,V. and Krishan,A. (1982) *Cancer Res.,* **42**, 1005.
53. Nederman,T. and Twentyman,P. (1984) in *Culture of Cellular Spheroids: A Handbook for Culture Techniques and Tumour Biological Tests. Chapter 5. Recent Results in Cancer Research,* Acker,H., Carlsson,J., Durand,R. and Sutherland,S.M. (eds.), in press.
54. Jones,A.C,. Stratford,I., Wilson,P.A. and Peckham,M.J. (1982) *Br. J. Cancer,* **46**, 870.
55. Kolb,A.I. (1981) *Lab. Equipment Digest,* **19**, 87.
56 Moon,T.E. (1980) in *Cloning of Human Tumor Stem Cells,* Salmon,S.E. (ed.), Alan R. Liss, Inc, NY, p. 209.
57. Berenbaum,M.C. (1974) *Lancet,* **ii**, 1141.
58. Meyskens,F.L.,Jr., Thomson,S.P., Hickie,R.A. and Sipes,N.J. (1983) *Br. J. Cancer,* **48**, 863.
59. Khan,W.H. and Ommaya,A.K. (1981) Abstract for *3rd International Symposium on Rapid Methods and Automation in Microbiology.*
60. International Conference: *Predictive Drug Testing on Human Tumour Cells,* Zurich, 20 – 22 July 1983.

216

Autoradiographic Localisation of Specific Cellular RNA Sequences Hybridised in Situ to Tritium-labelled Probes

D.CONKIE

1. INTRODUCTION

1.1 Principle

In the molecular hybridisation technique devised by Gillespie and Spiegelman (1) single-stranded nucleic acids in solution anneal to complementary regions of nucleic acid bound to nitrocellulose filters. In analogous experiments, double-stranded hybrid molecules form when exogenous single-stranded highly radioactive nucleic acid sequences anneal to complementary regions of nucleic acid fixed in cytological preparations of chromosomes (i.e., when hybridised *in situ*) (2,3). By hybridising to fixed preparations of cells the technique has been modified to permit detection of cytoplasmic RNA transcripts within individual cells (4,5).

1.2 Advantages of Hybridisation in Situ

Conventional techniques of hybridisation *in vitro* give an average value for the total target tissue. This complicates the interpretation of results obtained from populations of diverse cell types, or from a population of a single cell type which is non-uniform for some reason, such as stage of differentiation, or position in the cell cycle during asynchronous proliferation. The *in situ* hybridisation technique overcomes the problem, to some extent, since results relate to single cells or even to single chromosomes. The combination of hybridisation with cytological discrimination also provides information about the localisation of the molecular hybrid within the structure of the target tissue. Furthermore, the technique is applicable where only very small numbers of cells are available.

2. PRINCIPAL REQUIREMENTS

Molecular hybridisation *in situ* is governed by: (i) target nucleotide sequence redundancy (i.e., the number of specific nucleic acid sequences present in the cytological preparation); (ii) the purity of the probe; (iii) the specific activity of the radioactively labelled probe; (iv) the sensitivity of the hybridisation method; (v) signal enhancement; and (vi) the efficiency of the autoradiographic detection procedure.

2.1 Sequence Redundancy

Originally the technique was used most successfully where the nucleic acid to be detected was abundantly present as many copies in each cell. For example, the reiterated sequences of satellite DNA, amplified ribosomal DNA of amphibian oocytes, the laterally amplified DNA of dipteran larval polytene chromosomes, ribosomal and 5S RNA have been successfully investigated by hybridisation *in situ*. Furthermore, gene transcripts accumulating in the cell have been identified and localised by a similar technique.

By contrast, gene sequences present in only a single copy of mammalian cells have only more recently been reliably localised to a specific genomic site due mainly to the availability of high specific activity probes and the development of methods of signal enhancement.

2.2 Probe Purity

The results of molecular hybridisation *in situ* can only be meaningful if the probe used is sufficiently pure to permit specific hybridisation to a nucleic acid site since it is essentially a probe excess method, so that even minor impurities may hybridise. Whereas standard physical methods can produce probes of high purity from abundant nucleic acids such as ribosomal RNA, the relatively low concentrations of most mRNA species available by these methods are difficult to purify to homogeneity in acceptable quantity. However, recombinant DNA cloning methods have essentially eliminated the problem of contaminating nucleic acid sequences and in principle, can be used to prepare probes for analysing any gene or genomic site of interest.

2.3 Probe Specific Activity

The specific activity of a radioactive probe must be high enough to permit detection after hybridisation, within a reasonable autoradiographic exposure time and with a significant signal to background ratio. Exogenous nucleic acid probes may be RNA or DNA complementary to the cellular sequences of interest. Radioactive probes may be prepared by *in vitro* transcription with RNA polymerase, reverse transcriptase or by nick translation in the presence of ^{35}S- or tritium-labelled precursors. An RNA polymerase system, nick translation and reverse transcriptase kits containing the essential reagents with a detailed protocol for the preparation of isotopically labelled RNA or DNA probes are available from Amersham International plc or New England Nuclear Division of DuPont. It is also possible to label nucleic acids by iodination with ^{125}iodine. ^{3}H- or ^{125}I-labelling of identical RNA probes give qualitatively identical patterns of grain distribution after hybridisation *in situ* (6) and iodinated ribosomal RNA produces a quantitative correlation of grain counts when hybridised to meiotic cells of different DNA content (7).

Using conventional hybridisation *in situ* and probe labelling with tritium to provide a specific activity of 10^{8} d.p.m./μg, it can be calculated that a unique gene of about 10^{3} base pairs would generate one silver grain after 100 days of autoradiograph exposure. This is unacceptably long and in addition, results in the complication of distinguishing the signal from the background level.

Incorporation of a single ^{35}S-nucleotide precursor results in a specific activity of up to 10^9 d.p.m./μg by reverse transcription or 3 x 10^8 d.p.m./μg by nick translation. These activities are commensurately higher with two or three labelled precursors, although the incorporation of four ^{35}S-labelled precursors is inhibitory both to reverse transcription and to nick translation. Specific activities of up to 10^9 d.p.m./μg or better can now be obtained by nick translation using essentially carrier-free ^{125}I-labelled deoxycytidine (dCTP) (>1500 Ci/mmol, Amersham or NEN).

2.4 Sensitivity

Detection of multiple copies of a gene transcript in the cytoplasm of fixed cells is relatively independent of probe specific activity considerations applicable to detection of single-copy genes. However, other considerations apply. Thus, the sensitivity of the hybridisation method in localising specific cellular mRNA transcripts principally depends upon the degree of cellular RNA retention during pre-hybridisation procedures and the accessibility of the target RNA species.

2.4.1 *Retention of the Target Cellular RNA*

RNA retention in fixed cytological preparations has been monitored by acridine orange staining after each pretreatment step prior to hybridisation (8). The results suggest that fixation in 2% paraformaldehyde provides more reproducible retention of RNA than the ethanol/acetic acid fixative specified in the original protocols. Excellent RNA retention is obtained after fixation in methanol/trichloroacetic acid (9). Glutaraldehyde at 0.2% in 0.1 M sodium cacodylate, used as a fixative for cryosections provides a stable retention of RNA.

2.4.2 *Accessibility of Cellular RNA Sequences*

The accessibility of target cellular RNA sequences is optimal after incubation of cells or tissue sections in pronase or in proteinase K. Immediately following this digestion the cells are incubated in paraformaldehyde to cross link the exposed cellular RNA to the remaining cell components and to the gelatin coating of the slide (10). This digestion and paraformaldehyde fixation results in a 2-fold increase in detection sensitivity without impairing cellular morphology.

2.5 Signal Enhancement

10% dextran sulphate accelerates the hybridisation rate of DNA probes to immobolised nucleic acids and furthermore promotes the formation of probe networks when the DNA is initially randomly cleaved and single stranded (11). For example when DNA is radiolabelled by nick translation (12) and subsequently denatured the partially complementary DNA fragments reanneal to each other to form extensive DNA networks. Furthermore when preparing DNA probes replicated in plasmids it is advantageous to label the entire recombinant molecule since attached vector DNA can participate in network formation. It has been estimated that 10% dextran sulphate in the hybridisation buffer results in a further 4-fold increase in signal detection sensitivity.

Probe networks can also be constructed from poly(A)$^+$ probes which, following hybridisation, are then detected by [^{125}I]poly BrdUrd tailed DNA complex such as high molecular weight sea urchin DNA. The interspersed repeated sequences form a DNA network (13).

2.6 Autoradiographic efficiency

The efficiency of autoradiography in detecting a tritium labelled molecule is assumed to be around 5 – 10%. ^{125}I- and ^{35}S-labelled nucleotides provide a higher efficiency of grain development relative to ^{3}H disintegrations while retaining adequate light microscope resolution for many purposes, such as detection of cellular virus genes and cytoplasmic mRNA. ^{125}I-Labelled SV40 cRNA produces a significant number of grains in an autoradiographic exposure time of about one week when hybridised to detect single viral integration sites on diploid chromosomes (14).

3. PREPARATION OF REAGENTS AND MATERIALS

3.1 Materials

16 mm diameter coverslips	Soak overnight in chromic acid, wash in running tap water for 2 h then in distilled water. Store in 70% ethanol.
Gelatin-coated slides	Acid wash slides as above. From distilled water, dip slides once at room temperature in gelatin chrome alum freshly prepared. Allow to drain and dry in a dust free atmosphere and store at 4°C.

3.2 General Reagents

Gelatin chrome alum	Gelatin	5 g
	Chrome alum	0.5 g
	Distilled water	1 l
Phosphate-buffered saline (PBS)	Sodium chloride	8.0 g
	Potassium chloride	0.2 g
	Disodium hydrogen phosphate	1.15 g
	Potassium dihydrogen phosphate	0.2 g
	Distilled water	1 l
	Adjust to pH 7.3	
20 × Standard saline citrate (SSC)	Sodium chloride	175 g
	Sodium citrate	88 g
	Distilled water	1 l
	Adjust to pH 7.0	

Methanol

10% Trichloroacetic acid

3.3 Hybridisation Reagents

Rubber Solution	Dilute to a working consistency with

(Cow Proofings Ltd., Slough, Berks).	about 1/10th volume of $60-80°$ petroleum ether.
Buffer	$3 \times$ SSC containing 40% formamide (Fluka) adjusted to pH 6.5.
Probe	Dissolve the labelled nucleic acid in the buffer at about 5×10^7 d.p.m./ml. If double stranded, denature at 70°C for 10 min immediately before use.

3.4 Autoradiography

Kodak AR10 stripping film
Ilford K-5D or L4 liquid emulsion
Kodak D-19 developer
Ilford Hypam fixer with hardener
Sorensen's phosphate buffer, 67 mM

pH 6.9	Potassium dihydrogen phosphate	4.53 g
	Disodium hydrogen phosphate	5.93 g
	Distilled water	1 l
pH 5.75	Potassium dihydrogen phosphate	0.89 g
	Disodium hydrogen phosphate	8.38 g
	Distilled water	1 l

DPX neutral mountant (R.A.Lamb)
May Grunwald stain (Gurr)
Giemsa stain (Gurr)
Haematoxylin (Gurr)
Tartrazine (Gurr)

4. DETAILED PROTOCOL: HYBRIDISATION IN SITU

The detection of cellular RNA transcripts by hybridisation *in situ* is readily accomplished without the necessity for high specific activity probes or signal enhancement. Therefore a simplified protocol is presented providing the maximum opportunity for success.

4.1 Hybridisation in Situ

(i)   Monolayer cultures may be grown on glass or Thermanox (Lux, Flow Laboratories) coverslips. Cytological preparations may be prepared from suspension cultures by centrifuging cells on to a gelatin-coated glass slide (Shandon, Cytocentrifuge or Damon/IEC Cytobuckets). Thin cryostat sections of tissue or cell pellets may also be used.

(ii) Fix for 5 min in methanol followed by three washes in 10% trichloroacetic acid at 4°C.

(iii) Place in 70% ethanol and then air dry. Coverslip cultures should be attached to slides, cells uppermost, in a spot of DPX. These specimens can be stored for several weeks at 4°C if desired.

221

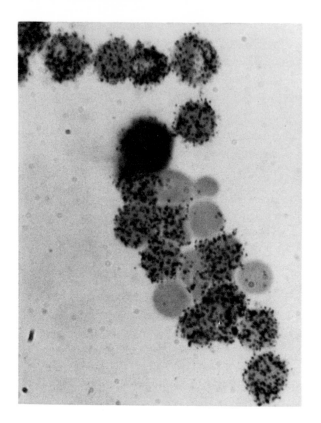

Figure 1. Rat globin cDNA hybridised to the peripheral blood cells of an anaemic rat (70% reticulocytosis as determined by supravital staining). The probe hybridised extensively to the majority of the anucleate cells whereas the lymphocytes and 28% of the anucleate cells are unlabelled. Hybridisation was to rat globin cDNA (prepared by Elke Korge. Sp. Act. 15×10^6 c.p.m./μg) in 3 x SSC/40% formamide at 43°C for 18 h. Post-hybridisation treatment, 0.5 h in 2 x SSC at 55°C. Detection of hybrid, Kodak AR10 stripping film autoradiography, 4-week exposure, developed in Kodak D19 diluted with an equal volume of water, fixed in Ilford Hypam. Staining, May-Grünwald 25 min, and Giemsa 25 min. Photomicrography, Ilford Pan F 35 mm film.

(iv)   Place 5 μl of the labelled probe solution directly onto the fixed cells previously warmed to 43°C by placing the slide on a metal tray floating in a water bath at 43°C.

(v)   Cover the droplet with a 16 mm diameter coverslip.

(vi)   Seal the edges of the coverslip with rubber solution.

(vii) When dry, check visually for a complete seal then incubate the slides for 18 h on the flat metal tray at 43°C in a water bath with lid. The high humidity is useful in the event of the gum seal being incomplete.

(viii) At the end of the incubation period cool on ice and peel the rubber seal from the coverslip using fine forceps.

(ix)   Remove the coverslips under the surface of 2 × SSC or, as an alternative, dislodge the coverslips with a jet of 2 × SSC.

(x)   Load slides into a glass slide carrier, immerse in 2 × SSC in a glass trough and seal the lid with tape.
(xi)   Partially immerse the trough in a 55°C water bath for 30 min.
(xii)   Wash the slides four times in 2 × SSC at 18°C.
(xiii) Dehydrate through an alcohol series up to absolute ethanol then air dry.
(xiv)   Detect molecular hybrids by autoradiography using stripping film or liquid emulsion.

The conditions of hybridisation and subsequent thermal stress should provide sufficient stringency to avoid non-specific binding of the labelled probe.

5. AUTORADIOGRAPHY

Photographic emulsion can be applied to the cells on a slide either by coating the slide with a strip of gelatin-backed emulsion or by dipping the slide into liquid emulsion.

Stripping film, AR-10 (Kodak), is a preformed layer of gelatin attached to a glass plate. When required for autoradiography the film is readily stripped from the glass plate in red safelight illumination. Only one crystal size, and so one sensitivity, is available. The gelatin supporting layer makes staining of the cells slightly more difficult following the autoradiographic procedures than when using liquid emulsion.

Dipping involves melting the nuclear emulsion in a water bath. Ilford K-5D, which requires no dilution, can be used in yellow-brown safelight illumination. Slides dipped in the emulsion pick up a thin layer over the specimen. Autoradiographs are relatively easy to stain since the gelatin of the emulsion forms a very thin layer. Emulsions are available with a range of grain size and sensitivity. However, it is more difficult to produce an emulsion layer of reproducible and uniform thickness than with stripping film.

5.1 Kodak AR 10 Stripping Film Autoradiography

(i)   Using a scalpel cut the AR 10 stripping film into strips of approximately 40 × 25 mm.
(ii)   Peel the strips from the glass backing plate and place on the surface of distilled water at 26°C, emulsion side down, for 3 min to allow the emulsion to swell by hydration.
(iii)   Place the slide with specimen uppermost into the water and under the film. Lift up under the film so that the film is in close contact with the specimen and drapes around the slide.
(iv)   After air-drying, expose the film at 4°C in a light-proof box containing silica gel.
(v)   After an appropriate exposure time develop the film by immersing the slide for 5 min at 18°C in Kodak D19 developer diluted with an equal volume of distilled water.
(vi)   Without washing, fix the autoradiographs by immersion for 3 min at 18°C in Ilford Hypam fixer containing acid hardener.
(vii)   Rinse in tap water and transfer to hypo cleaving agent (Kodak) for 2 min.
(viii) Wash gently for 5 min in cold running water at 10−18°C then air dry.

5.2 Staining Stripping Film Autoradiographs

(i)   Stain the autoradiographs for 25 min in May-Grünwald stain diluted with an equal volume of Sörensen's 67 mM phosphate buffer pH 5.75 at 18°C. To prevent the deposition of any precipitate which might form as the stain is gradually oxidised, the slides should be immersed and stained vertically. To avoid contamination by the oxidised scum which forms on the surface of the stain, slides should not be withdrawn at the end of the staining period. It is preferable to fill the staining dish until overflowing by addition of excess buffer at 18°C so that the scum floods over the top of the dish.

(ii)   Slides can then be withdrawn and stained in 5% Giemsa (Gurr) in buffer at pH 5.75, 25 min at 18°C. The same precautions should be observed as for the May-Grünwald stain.

(iii)   Wash the slides gently in cold water, air dry and mount in DPX.

(iv)   Alternatively, autoradiographs may be stained in haematoxylin for 2 min followed by a further 15 min in gently running tap water at 10−18°C. The autoradiographs may then be counterstained in tartrazine for 5 min, washed, dried and mounted in DPX.

5.3 Liquid Emulsion Autoradiography

(See also Chapter 7)

(i)   Coat slides with emulsion melted at 50°C (and diluted with a double volume of distilled water at 50°C if required by manufacturer's detailed instructions).

(ii)   Drain slides and wipe the backs free of emulsion.

(iii)   Air dry and place in a light-tight box at 4°C. To avoid rapid dehydration of the thin emulsion layer and consequent cracking, it is advisable to delay adding silica gel for 24 h. Expose for required time.

(iv)   Process the film as before.

5.4 Staining Liquid Emulsion Autoradiographs

(i)   Stain for 15 min in May-Grünwald stain diluted with a double volume of Sörensen's buffer pH 6.9 at 18°C.

(ii)   Wash in buffer and stain in 5% Giemsa pH 6.9 at 18°C for 15 min. The methods for avoiding stain deposit should be observed.

(iii)   Wash gently in cold water, air-dry and mount in DPX.

(iv)   Alternatively, stain with haematoxylin and tartrazine as before.

   Further information on autoradiography is available in a review available from Amersham International (15).

6. THE DETECTION OF MOLECULAR HYBRIDS IN CYTOLOGICAL PREPARATIONS BY NON-RADIOACTIVE METHODS

Alternative methods for the detection of cellular transcripts hybridised *in situ* using fluorescent or enzyme-conjugated labels have some advantages over conventional autoradiographic detection procedures. For example: (i) the time required to determine the site of hybridisation is decreased markedly by eliminating autoradiographic exposure; (ii) fluorescence detection methods have a resolving power equal to or greater than that achieved autoradiographically; and (iii) a lower background noise level is often obtainable.

The simplest of these alternative detection procedures employs partially depurinated DNA probes bound to acriflavin (16). This can be detected by fluorescence after hybridisation *in situ*. Rudkin and Stollar (17) describe the use of a specific immune serum raised in rabbits against double-stranded DNA-RNA complexed with methylated BSA. The complex is then detected *in situ* by goat anti-rabbit IgG conjugated to rhodamine. Using biotin covalently attached to a nucleic acid probe, Manning *et al.* (18) have devised a novel detection system. The biotin-RNA-DNA hybridisation site can be labelled with polymethacrylate spheres covalently coupled to avidin due to the strong interaction between biotin and avidin. The spheres are identified by scanning electron microscopy and sensitivity is proportional to the number of spheres binding at the site of nucleic acid hybrid molecules. A more sensitive detection system has been described by Cheung *et al.* (19), using highly fluorescent latex microspheres conjugated with polyuridylic acid. The latex microspheres are annealed with the 3' end of the probe cRNA after hybridisation *in situ*. The resultant double hybrid is visualised by conventional fluorescence microscopy. More recently, Langer *et al.* (20) have synthesised nucleotide analogues of deoxyuridine or uridine triphosphates containing a biotin molecule covalently attached to the C-5 position of the pyrmidine ring through 'linker' allylamine groups. The biotinylated analogue of deoxyuridine triphosphate (dUTP) can be incorporated into DNA probes by nick translation. Such polynucleotides have been used in the *in situ* hybridisation technique with superior resolution and signal-to-noise ratio (21). Methods for detecting this probe after hybridisation *in situ* are either: (a) immunological using anti-biotin antibodies in consort with a second antibody tagged with a fluorescent, enzymatic or electron-dense reagent; or (b) affinity labelling using preformed complexes of avidin conjugated to rhodamine or avidin with biotinylated derivatives of peroxidase or alkaline phosphatase. Biotinylated dUTP for use in nick translations, together with avidin and biotinylated polyalkaline phosphatase are available as a DNA detection system from Bethesda Research Laboratories.

7. ACKNOWLEDGEMENT

Original work described is supported by grants from the Cancer Research Campaign.

8. REFERENCES

1. Gillespie,D. and Spiegelman,S. (1965) *J. Mol. Biol.*, **12**, 829.
2. John,H.A., Birnstiel,M.L. and Jones,K.W. (1969) *Nature*, **223**, 582.
3. Gall,J.C. and Pardue,M.L. (1969) *Proc. Natl. Acad. Sci. USA*, **63**, 378.
4. Harrison,P.R., Conkie,D., Paul,J. and Jones,K. (1973) *FEBS Lett.*, **32**, 109.
5. Conkie,D., Affara,N., Harrison,P.R., Paul,J. and Jones,K. (1974) *J. Cell Biol.*, **63**, 414.
6. Attenburg,L.C., Getz,M.J., Crain,W.R., Saunders,G.F. and Shaw,M.W. (1973) *Proc. Natl. Acad. Sci. USA*, **70**, 1536.
7. Wolgemuth,D.J., Jagello,G.M., Atwood,K.C. and Henderson,A.S. (1976) *Cytogenet. Cell Genet.*, **17**, 137.
8. Hafen,E., Levine,M., Garber,R.L. and Gehring,W.J. (1983) *EMBO J.*, **2**, 617.
9. Birnie,G.D., Burns,J.H., Clark,P. and Warnock,A.M. (1984) *J. R. Soc. Med.*, **77**, 289.
10. Singer,R.H. and Ward,D.C. (1982) *Proc. Natl. Acad. Sci. USA*, **79**, 7331.
11. Wahl,G.M., Stern,M. and Stark,G.R. (1979) *Proc. Natl. Acad. Sci. USA*, **76**, 3683.

12. Rigby,P.W.J., Dieckmann,M., Rhodes,C. and Berg,P. (1977) *J. Mol. Biol.*, **113**, 237.
13. Tereba,A., Lai,M.M.C. and Murti,K.C. (1979) *Proc. Natl. Acad. Sci. USA*, **76**, 6486.
14. Szabo,P. and Ward,D.C. (1982) *Trends Biochem. Sci.*, **7**, 425.
15. Rogers,A.W. (1979) Practical autoradiography, available as Review 20 from Amersham International.
16. Levinson,J.W., Desostoa,A., Liebes,L.F. and McCormick,J.J. (1976) *Biochim. Biophys. Acta*, **447**, 260.
17. Rudkin,G. and Stoller,B. (1977) *Nature*, **265**, 472.
18. Manning,J.E., Hershey,N.D., Broker,T.R., Pellegrini,M., Mitchell,H.K. and Davidson,N. (1975) *Chromosoma*, **53**, 107.
19. Cheung,S.W., Tishler,P.V., Atkins,L., Sengupta,S.K., Modest,E.J. and Forget,B.G. (1977) *Cell Biol. Intl. Rep.*, **1**, 255.
20. Langer,P.R., Waldrop,A.A. and Ward,D.C. (1981) *Proc. Natl. Acad. Sci. USA*, **78**, 6633.
21. Langer-Safer,P.R., Levine,M. and Ward,D.C. (1982) *Proc. Natl. Acad. Sci. USA*, **79**, 4381.

APPENDIX I

General Reagents

Hanks' Balanced Salt Solution (HBSS) (Modified)

$CaCl_2$	0.14 g	(1.26 mM)
KCl	0.40 g	(5.36 mM)
KH_2PO_4	0.06 g	(0.44 mM)
$MgCl_2$ $6H_2O$	0.10 g	(0.49 mM)
$MgSO_4$ $7H_2O$	0.10 g	(0.41 mM)
NaCl	8.00 g	(0.137 M)
Na_2HPO_4 $7H_2O$	0.09 g	(0.34 mM)
Phenol Red	0.01 g	
Water to	1000 ml	

Autoclave below pH 6.5, 15 p.s.i. (1 bar), 20 min.
Before use pH may be adjusted to 7.4 with sterile NaOH and sterile 1 M Hepes may be added to 20 mM as a buffer.
Glucose, if required, should be autoclaved separately at 100 g/l and diluted 1:100 to give 1.0 g/l (5.55×10^{-3} M).

This recipe is provided as a general handling, washing and dissection solution. If used as a base for growth medium, glucose (see above) and 0.35 g/l $NaHCO_3$ (4 mM) must be added. It is then suitable for culture in a sealed container with air as a gas phase.

Dulbecco's Phosphate-buffered Saline, Solution A (PBSA, Ca^{2+} and Mg^{2+} free)

KCl	0.20 g	(2.68 mM)
KH_2PO_4	0.20 g	(1.47 mM)
NaCl	8.00 g	(0.137 M)
Na_2HPO_4 $7H_2O$	2.16 g	(8.06 mM)
Water to	1000 ml	

Sterilise by autoclaving 15 p.s.i. (1 bar) 20 min.
Used as a washing solution before disaggregation, and as a diluent for trypsin.

Dissection BSS (DBSS)

HBSS or PBSA with 250 u/ml penicillin, 250 μg/ml streptomycin, 100 μg/ml kanamycin or 50 μg/ml gentamycin, and 2.5 μg/ml fungizone.

PBSA/EDTA

Na_2EDTA $2H_2O$	0.372 g	(1 mM)
PBSA to	1000 ml	

Sterilise by autoclaving 15 p.s.i. (1 bar) 20 min.

Trypsin

2.5 g (crude Difco 1/250) or 0.1 g (3 $\times$ recrystallised, Sigma).
HBSS or PBSA or PBSA/EDTA to 1000 ml, as required to obtain complete disaggregation.

APPENDIX II

Suppliers of Specialist Items

Amersham International, White Lion Road, Amersham, Bucks HP7 9LL, UK

Amicon Ltd, Amicon House, 2 Kingsway, Woking, Surrey GU21 1UR, UK

Arbrook Ltd., Livingstone, Midlothian, UK

Artek Systems Corp., 170 Finn Court, Farmingdale, NY 11735, USA

Baird and Tatlock (London) Ltd., PO Box 1, Romford, Essex RM1 1HA, UK

BDH Chemicals Ltd., Poole, Dorset BH12 4NN, UK

Beckman-RIIC Ltd., Analytical Instruments Sales and Service Operation, Cressex Industrial Estate, Turnpike Road, High Wycombe, Bucks, UK

Becton Dickinson (UK) Ltd., Between Towns Road, Cowley, Oxford OX4 3LY, UK

Bellco (see A R Horwell for UK)

Bellco Glass Inc., 340 Edrudo Road, Vineland, NJ 08360, USA

Biomedical Technologies Inc., 22 Thomdike St., Cambridge, MA 02141, USA

Biotec Instruments Ltd., Unit 1, Caxton Hill Extn. Rd., Caxton Hill, Hertford SG13 7LS, UK

Boehringer Mannheim GmbH, Biochemica, Postfach 310120, D-6800 Mannheim 31, FRG

BRL, Bethesda Research Laboratories GmbH, Offenbacher Strasse 113, D-6078 Neu-Isenburg 1, FRG

Brooks, J Laboratories, Products Serving Biotechnology, PO Box 37, Olivenhain, CA 92024, USA

Browne, Albert Ltd., Chancery Street, Leicester LE1 5NA, UK

Chemap, Div. A C Biotechnics Inc., 230 Crossways Park Drive, Woodbury, NY 11746, USA

Cole-Farmer Instrument Co., 7425 N. Oak Park Ave., Chicago, IL 60648, USA

Corning Ltd., Stone, Staffs ST15 0BG, UK

Coulter Electronics Ltd., Northwell Drive, Luton, Beds LU3 3RH, UK

CR, Research Products Division, Collaborative Research Inc., 128 Spring St., Lexington, MA 02173, USA and 1365 Main St., Waltham, MA 02154, USA

Cryo-Med, 49659 Leona Drive, Mt. Clements, MI 48043, USA

Cryoservice Ltd., Platts Common Industrial Estate, Hawshaw Lane, Hoyland, Barnsley, Yorks, UK

Dako Ltd., 22 The Arcade, The Octagon, High Wycombe, Bucks HP11 2HT, UK

Decon Laboratories Ltd., Conway St., Hove, E. Sussex BN3 3LY, UK

Dextran Products Ltd., PO Box 1360, Princeton, NJ 08542, USA

Dow Chemical Co., Heathrow House, Bath Road, Hounslow TW5 9QY, UK

Du Pont UK Ltd., 2 New Road, Southampton SO2 0AA, UK

eba, European Business Associates, 68 rue de la Petrusse, L-8084 Betrange, Luxembourg

Elga Group, Lane End, High Wycombe, Bucks, UK

Expanded Metal Co., Hartlepool, UK

Flow Laboratories Ltd., PO Box 17, Second Ave. Ind. Estate, Irvine, Ayrshire KA12 8NB, UK

FMC Corporation, Rockland, Maine, USA

Gelman Sciences Inc., 600 S. Wagner Road, Ann Arbor, MI 48106, USA (see also Flow Labs. for UK)

Genzyme, IC Chemikalien GmbH, Sohnckestrasse 17, D-8000 München 71, FRG

Gibco Europe Ltd., PO Box 35, Trident House, Renfrew Road, Paisley PA3 4EF, UK

Glaxo Laboratories Ltd., Greenford, Middlesex, UK

Gurr (see BDH)

Hana Media Inc., A subsidiary of Hana Biologics Inc., 629 Bancroft Way, Berkeley, CA 94710, USA

Heraeus, W C GmbH PEW, Postfach 1220, D-3360 Osterade am Harz, FRG

Hoechst (UK) Ltd., Hoechst House, Salisbury Road, Hounslow, Middlesex TW4 6JH, UK

Hopkin and Williams, PO Box 1, Romford, Essex RM1 1HA, UK

K C Biologicals Europe, 11 Chemin de Ronde, F-78110 Le Vesinet, France

K C Biological, Subsidiary of Corning Glass Works, PO Box 14848, Lenexa, KS 66215, USA

Kodak Ltd., PO Box 66, Kodak House, Station Road, Hemel Hempstead, Herts HP IJU, UK

KOR Biochemicals, 69 rue de la Petrusse, L-8084 Betrange, CD Luxembourg

L H Fermentation, Bells Hill, Stoke Poges, Slough, Bucks SL2 4EJ, UK

L'Aire Liquide, Paris, France

L'Aire Liquide (UK) Ltd., 44 Hertford St., London W1Y 7TF, UK

Leitz (E) (Instruments) Ltd., Luton, Beds., UK

LKB, The Incentive Group, LKB-Produkter AB, Box 305, S-16126 Bromma, Sweden

Millipore (UK) Ltd., Millipore House, 11-15 Peterborough Rd., Harrow, Middlesex HA1 2BR, UK

Morgan Sheet Metal Co, Sarasota, FL, USA

NEN Research Products, 549 Albany St., Boston, MA 02118, USA

New Brunswick Scientific, 26-34 Emerald St., London WC1N 3QA, UK

Northumbria Biologicals, S. Nelson Industrial Estate, Cranlington, Northumberland NE23 9HL, UK

Nucleopore Corp., 7035 Commerce Circle, Pleasanton, CA 94566, USA

Nunc (see Gibco Europe Ltd.)

Ortho Diagnostic Systems Ltd. (Division of Ortho Pharmaceuticals Ltd.), Enterprise House, Station Road, Loudwater, High Wycombe, Bucks, UK

Ortho Diagnostic Systems Inc., 410 University Avenue, Westwood, MA 02090, USA

Paesel GmbH and Co., Biochemika, Diagnostika, Pharmazeutika, Borsigallee 6, D-6000 Frankfurt/Main 63, FRG

Pall Process Filtration Ltd., Europa House, PO Box 62, Portsmouth, Hants PO1 3PD, UK

Pharmacia (Great Britain) Ltd., Prince Regent Road, Hounslow, Middlesex TW3 1NE, UK

Planer Products Ltd., Windmill Road, Sunbury on Thames, Middlesex TW16 7HD, UK

Polysciences Ltd., 24 Low Farm Place, Moulton Park, Northampton NN3 1HY, UK

Reactifs IBF, Societe Chimique Pointet-Girard, 35 avenue Jean-Jaures, F-92390 Villeneuve-La-Garenne, France

Schering Chemicals Ltd., Pharmacuetical Division, Burgess Hill, Sussex RH15 9NE, UK

Sera Lab., Crawley Down, Sussex RH10 4LL, UK

Seragen Inc., 54 Clayton St., Boston, MA 021122, USA

Sigma Chemical Co. Ltd., Fancy Road, Poole, Dorset BH17 7NH, UK

Squibb and Sons Ltd., Regal House, Twickenham, Middlesex TW1 3QT, UK

Sterilin Ltd., 43-45 Broad St., Teddington, Middlesex TW11 8QZ, UK

Union Carbide UK Ltd., Cryogenics Division, Redworth Way, Aycliffe Industrial Estate, Aycliffe, Co. Durham, UK

Usher, Dextran Products Ltd., 421 Comstock Road, Scarborough, Ontario, Canada M1L 2H5

Ventrex, 217 Read St., Portland, ME 04103, USA

Watson-Marlowe Ltd., Falmouth, Cornwall TR11 4RU, UK

INDEX

Colony stimulating factors,
role in mammalian cell culture 13
Commercially available media,
serum-additive 28
serum-free media 28
serum-substitute 28
Complementary sequences,
probe specific activity 218
Computer,
flow cytometry, design of equipment 126
Confluence,
growth kinetics 10
Contamination,
determination 10
reduced risk using serum-free media 20
Continuous cell line,
definition 5
Continuous cell lines,
normal/tumorigenic 4
Continuous-flow culture,
apparatus 66
chemostat 65
growth-limiting nutrient 65
use of microcarriers 45
Courtenay method,
suspension cloning 202
Cross contamination,
cell lines 71
determination 10
Cryoprotective agents,
dimethyl sulphoxide/glycerol 73
Cryostat sections,
hybridisation in situ 220
Culture vessels 7
scale-up 35
Cycle-specific drugs,
pharmacokinetics, drug exposure 189
Cyclophosphamide,
effects on radioisotope incorporation,
cytotoxicity assays 194
microsomal enzyme activation 199
Cytochrome-P450,
drug activation 200
Cytodex,
microcarriers 54
Cytodifferentiation,
embryological watchglass technique 159
Cytofluorimetry,
definition 125
CYTOFLUOROGRAF,
description 127
Cytogenetics,
species verification 78,82
Cytogram,
cell sorting 130

Cytograms,
CYTOFLUOROGRAF 129
data output 134
Cytokeratin,
markers, tissue of origin verification 99
Cytoskeleton,
intermediate filament proteins, cell
identification 101
Cytotoxic activity,
anti-metabolites 199
Cytotoxicity,
assays 183
comparison with anti-neoplastic activity
213
interpretation of results 192
Cytotoxicity assays,
protein determination 205
Cytotoxicity index,
interpretation of results 209

Data banks 109
CODATA-IUIS hybridoma data bank
110
MIRDAB 111
Data output,
flow cytometry 134
Dehydrogenase,
respiration/glycolysis, drug-induced
changes 194
Deionised water,
use for glassware washing 8
Delayed cytotoxicity,
recovery period 191
suspension cultures 186
Desmosomes,
markers, tissue of origin verification 99
Development,
autoradiography 223
Dexamethasone,
role in mammalian cell culture 15
Dextran sulphate,
signal enhancement 219
Diastase,
exclusion of glycogen, periodic acid-
Schiff staining 174
Differentiation,
study of, grid techniques 167
Dihydrotestosterone,
testosterone metabolism, prostate glands
167
Dimethyl sulphoxide,
cryoprotective agents, cell freezing 73
Dipeptidase 1,
polymorphic isozymes, intraspecies
cross-contamination monitoring 95

Ribosomal DNA,
  hybridisation in situ 218
RNA polymerase,
  probe specific activity 218
RNA sequences,
  paraformaldehyde fixation 219
RNA staining,
  acridine orange 142
RNA synthesis,
  organ culture, autoradiography 177
  testosterone, prostate glands 167
Roller bottle,
  use in scale-up 48, 49

Safety evaluation,
  mutagenicity/carcinogenicity/teratogenici-
    ty 183
Sampling medium,
  scale-up 35
Satellite DNA,
  hybridisation in situ 218
Scale-up,
  anchorage-dependent cells 46
  equipment 35
  medium supplements 36
  oxygen limitation 43
  pH stabilisation 36
  suspension culture 59, 60
  use of hollow fibre culture 50
  use of multitray unit 50
  use of plate heat exchanger 51
  use of roller bottle 48, 49
  use of suspension culture 33
Secretory activity,
  determination, Maximow single slide
    technique 156
Seed stock,
  effect of phenotypic drift/senescence 72
Selective growth media,
  clonogenic assays 188
Selenite,
  serum-additive 28, 29
  serum-free media 25
Selenium,
  role in mammalian cell culture 15
Senescence,
  cell line banking 72
  effect on seed stock 72
Serum,
  attachment factors/spreading factors, role
    in mammalian cell culture 15
  role in mammalian cell culture 13
  use in growth media 6
Serum batch testing 6

Serum components,
  role in mammalian cell culture 13
Serum cytotoxicity,
  avoidance, use of serum-free media 20
Serum dependence,
  continuous cell line 5
Serum-free media 28
Serum-free media,
  advantages, mammalian cell culture 15
  mammalian cell culture 13
  requirements 6
Serum-substitute 28
Shear forces,
  effects of impellers 63
  problems in stirrer vessels 62
Signal display,
  fluorescence activated cell sorter 134
Signal enhancement,
  dextran sulphate/probe networks 219
Signal processing,
  fluorescence activated cell sorter 134
Skeletal tissues,
  grid techniques 151
Skin,
  explanation, organ culture 166
  organ culture 149
  organ culture, grid techniques 165
Sodium carboxymethyl cellulose,
  medium supplements, scale-up 36
Solid tumours,
  disaggregation 139
Solvent controls,
  drug formulation 197
  drug storage 198
Somatomedins,
  role in mammalian cell culture 14
Source of tissue,
  cell culture/organ culture 3
Sparging,
  oxygen limitation, microcarrier culture
    58
  oxygen transfer 42
Species verification,
  cytogenetics 78
  fluorescent antibody staining 78
  iso-enzymes 78,80
  karyology/chromosome analysis 82
Spheroids 2
Spheroids,
  cytoxicity assays 203
  drug penetration 185
  survival assays 196
Spinner cultures,
  commercially available models 62

Published
in the
Practical
Approach
series

Mutagenicity testing

a practical approach

Mutagenicity
testing

a practical approach

Edited by
S Venitt & J M Parry

Published in the
Practical Approach Series
Series editors: D Rickwood and B D Hames

○IRL PRESS
Oxford · Washington DC

Edited by
S Venitt, Institute of Cancer Research,
and
J M Parry, University College of Swansea

A laboratory manual of genetic toxicology

Mutagenicity testing: a practical approach describes in great practical detail nine different tests for detecting mutagenic activity, chromosomal aberrations and induction of DNA damage and repair.

The book is written for researchers in universities and industry, post-graduates and undergraduates in toxicology, industrial hygiene, medicine and biochemistry. The editors introduce the aims and concepts of genetic toxicology and nine chapters give practical instructions for conducting tests in organisms ranging from bacteria to laboratory rodents.

Particular emphasis is given to well-validated tests required by national and international regulatory agencies. The text also offers guidance on the presentation and interpretation of results and includes key references and advice on laboratory safety.

CONTENTS
Background to mutagenicity testing *S Venitt and J M Parry*

The analysis of alkylated DNA by high pressure liquid chromatography *W Warren*

Bacterial mutation assays using reverse mutation *S Venitt, C Crofton-Sleigh and R Forster*

DNA repair tests in cultured mammalian cells *R Waters*

The assay of the genotoxicity of chemicals using the budding yeast *Saccharomyces cerevisiae E M Parry and J M Parry*

Mutation tests with the fruit fly *Drosophila melanogaster I de G Mitchell and R Combes*

Assays for the detection of chemically-induced chromosome damage in cultured mammalian cells *B J Dean and N Danford*

The detection of gene mutations in cultured mammalian cells *J Cole and C F Arlett*

Cytogenetic tests in mammals *I-D Adler*

The dominant lethal test in rodents *D Anderson*

December 1984; 368 pp; 0 904147 72 X (softbound)

For details of price and ordering consult our current catalogue or contact:
IRL Press Ltd, PO Box 1, Eynsham, Oxford OX8 1JJ, UK
IRL Press Inc, PO Box Q, McLean, VA 22101, USA

◇IRL PRESS
Oxford · Washington DC

Published in the Practical Approach series

Plant cell culture
a practical approach

Edited by R A Dixon, *Royal Holloway College, London*

Plant cell culture

a practical approach

Edited by
R A Dixon

July 1985;
252pp;
0 947946 22 5
(softbound)

Different plant species respond differently to culturing and manipulation. *Plant cell culture* provides the knowledge necessary to rationalise these differing responses while, at the same time, offering effective laboratory-bench methods for dealing with particular cases.

Plant cells grown in aseptic culture offer exciting rewards for researchers in biotechnology and molecular biology. This new handbook departs from the agricultural emphasis of previous works to focus on plant cell culture as a successful and important tool for research.

The book describes how to establish cell cultures, the production and manipulation of protoplasts and the pathways of embryogenesis, organogenesis and regeneration. Other aspects described include vascular differentiation, selection for inhibitor resistance and the formation of secondary products. The book ends with plant – virus and plant – fungus interactions and protocols for cryopreservation of plant tissue.

Contents
Isolation and maintenance of callus and cell suspension cultures *R A Dixon* ● Haploid cell cultures *J M Dunwell* ● Isolation, culture and genetic manipulation of plant protoplasts *J B Power and J V Chapman* ● Selection of plant cells for desirable characteristics: inhibitor resistance *R A Gonzales and J M Widholm* ● Embryogenesis, organogenesis and plant regeneration *B Tisserat* ● Use of tissue cultures for studies on vascular differentiation *G P Bolwell* ● Secondary product formation by cell suspension cultures *P Morris, A H Scragg, N J Smart and A Stafford* ● Cryopreservation and storage of germplasm *L A Withers* ● Tissue culture methods in phytopathology I: viruses *K R Wood* ● Tissue culture methods in phytopathology II: fungi *S A Miller*

For details of price and ordering consult our current catalogue or contact:
IRL Press Ltd,
Box 1, Eynsham,
Oxford OX8 1JJ, UK

IRL Press Inc,
PO Box Q,
McLean VA 22101,
USA

IRL PRESS
Oxford · Washington DC